Cool Facts for Kids

Mind-Blowing Science, Epic History, Fun Pop Culture & Strange World Facts

WIZKID PUBLISHING

Table of Contents

Introduction

Did you know there are more trees on Earth than stars in our galaxy? Close your eyes and imagine walking through a forest, where each leafy giant carries stories and secrets, whispering tales of storms weathered and sunlight captured. Now open them wide to the stars, flickering far above us—each a distant world with its own mysteries. This book is your gateway to uncovering such wonders, one astonishing fact at a time.

Welcome, young explorers, to an incredible adventure that promises to take you across time and space, unraveling the mysteries of the universe and the marvels of human achievement. Whether it's the enigmatic allure of black holes, the melodious magic of classical music, or the intricate dance of numbers in mathematics, each chapter awaits like a hidden treasure chest, bursting with knowledge, mystery, and wonder.

At the heart of this thrilling expedition lies curiosity—the spark that ignites our thirst for understanding, driving us to peer deeper into the unknown. It is the same curiosity that sends astronauts soaring into the cosmos or inspires scientists to unlock the secrets of ancient fossils. Here, we celebrate that inquisitive spirit, encouraging you to let your imagination wander amidst these pages. Every question you ponder could be the beginning of your own journey to discovery.

Imagine standing at the threshold of an immense library, each shelf filled with books detailing everything from how gravity holds us safely to the ground to why certain melodies can evoke tears of joy or sorrow. This is no ordinary library but a portal to countless adventures awaiting your exploration. You'll discover that learning isn't just about memorizing facts—it's about embarking on journeys that teach you how to think, dream, and connect with the fascinating tapestry of the world around you.

Throughout this book, you'll delve into captivating chapters that invite you to explore various realms of knowledge. Discover how the stars have guided sailors across oceans, and how ingenious inventions continue to shape our modern lives. Uncover the natural wonders that make earth's ecosystems teem with life, and learn about historical events that shaped humanity's path through the ages. Each chapter is crafted to pique your interest and leave you eager for more.

But the adventure doesn't stop here. As you turn each page, let the stories inspire you to seek further—to dig deeper into subjects that excite you. Ask questions, start conversations, and share your newfound knowledge with friends and family. Your curiosity is a powerful tool, capable of opening new doors and leading you on paths you never imagined possible.

Imagine the thrill of discovering something entirely new—a fact that no one else knows. Perhaps your exploration of the stars will lead you to discover an uncharted planet or invent the technology of the future. Maybe the seeds of creativity planted by reading about the world's greatest artists will inspire your masterpiece—a painting, a song, or even a groundbreaking idea that changes the world.

The possibilities are endless when you fuel your mind with knowledge and imagination. This book aims not only to inform but also to awaken a lifelong love for learning that extends beyond textbooks and classrooms. It seeks to plant the seed of curiosity within you, nurturing it until it grows into a vast tree of wisdom and creativity.

As we begin this journey together, remember that every great explorer faces challenges. There will be times when you don't understand something right away, or when the path seems too winding. But that's all part of the adventure! Embrace those moments as opportunities to learn and grow. Every stumbling block is a stepping stone along the path of discovery.

Find excitement in the unexpected, and allow the thrill of learning to guide you through obstacles. Celebrate your successes and learn from your mistakes, for they often hold the keys to enlightenment. Be brave, be curious, and most importantly, enjoy every moment of this marvelous expedition through the pages of this book.

In closing, I invite you to dive headfirst into this kaleidoscope of knowledge letting your natural curiosity propel you forward. With each page, you'll uncover stories and facts that are meant to be shared—a perfect chance to become the storyteller among your friends, sparking conversations that connect and inspire.

And as you explore, imagine the limitless expanse of the universe outside your window. Keep asking questions, keep looking up at the stars, and dream of what lies beyond. Who knows? You might just be the one who writes the next chapter in the saga of human discovery.

So, let's begin, shall we? Welcome to a world of infinite wonders, waiting for their curtain to rise. The adventure beckons—let your journey commence!

Chapter 1

Fact or Fiction: The Truths of Science Uncovered

Unraveling the secrets of the universe is like solving a grand cosmic puzzle. These intriguing theories stir our curiosity and push the boundaries of what we consider possible, playing on the edge of scientific reality and wild dreams.

Let's explore some of the most captivating and revolutionary theories within science that rather seem the plot of a science fiction film!

The Mystery of Wormholes

Have you ever wondered about the possibility of traveling through space in mere seconds, as if by magic? Let me introduce you to wormholes, fascinating concepts that exist mainly within the theoretical realm of physics.

- Imagine these wormholes as tunnels or secret passages with the power to connect different points in our universe more quickly than any spaceship could fly.

- Wormholes come from the equations of general relativity, a framework created by the great physicist Albert Einstein.

- They are like shortcuts through space and time, potentially allowing someone to leap vast distances instantly.

- There are different types of wormholes:

- The Traversable wormholes, which could theoretically allow for travel, are like bridges connecting black holes to distant regions of space.

- The Einstein-Rosen bridges, named after Einstein and another physicist, Nathan Rosen, who first proposed them. Unfortunately, Einstein-Rosen bridges are not available for time travel in space because they collapse too quickly. Things wouldn't have enough time to pass through.

- To understand what a wormhole looks like, think of a worm tunneling through an apple. The worm doesn't crawl over the surface to reach the other side; it moves right through the fruit! In space, it happens the same: a wormhole creates a shortcut between faraway points in space and time.

- For travel through a traversable wormholescientists believe we need exotic matter.

- This exotic matter isn't the kind of stuff we encounter every day; it's purely theoretical and has an unusual quality called negative energy density. Think of it like having something that pushes against gravity instead of being pulled by it.

- Exotic matter is a bit mysterious because we haven't observed or created it yet, and it exists mostly in theories crafted by physicists.

- The idea of exotic matter is easier to grasp when you imagine a tunnel made of soft material that keeps collapsing under its own weight. To keep this tunnel open, you might insert an inflatable structure, pushing its walls outward.

- Exotic matter would act like this supportive structure, keeping the wormhole's throat from collapsing, allowing it to remain open for potential travel.

Even though concepts like wormholes and exotic matter stir the imagination, we don't currently have the technology to find or stabilize them. Such notions live on the edge of scientific possibility, leaving us curious about what might be out there in the universe.

- Scientists continue to explore and learn, hoping one day to unlock these tantalizing mysteries and turn what we imagine into reality.

- Some audacious scientists suggest wormholes could do more than connect distant areas in space. They might also allow for time travel!

- Imagine stepping into one end of a wormhole and emerging in the past or future.

Some Boos From the Audience

Even though scientists of all time have tried to travel in time, this fantastic journey is still a challenge to the laws of physics. Physicists haven't found a way to explain through concepts and formulas how this journey through wormholes would work. Moreover, some people say it is impossible...

- As intriguing as these possibilities are, not everyone agrees on the feasibility of using wormholes for travel.
- Some respected scientists point out that while wormholes theoretically fit within the math of general relativity, practical hurdles remain gigantic.
- Stephen Hawking, a famous cosmologist, argued that turning a wormhole into a time machine isn't really possible. Instead, he described a wormhole as a quick route, bringing something seemingly far away much closer.

Where Have We Heard About Wormholes?

In many movies and books, wormholes are interesting pathways that let characters travel to different times, face their pasts, or see possible futures.

- Interstellar
    - Set in a dystopian future where Earth is becoming uninhabitable, a group of astronauts, led by Cooper, travels through a wormhole near Saturn in search of a new habitable planet.
- Time Force
    - A team of time-traveling police officers from the future are sent back to the 21st century to capture a criminal who has escaped into the past.
- Miss Peregrine's Peculiar Children
    - After discovering a mysterious island and an orphanage, a young boy named Jacob learns about children with

unique abilities living under the protection of Miss Peregrine.

What If...?

If you found a wormhole in your backyard and discovered you could travel in time, would you go back or forth? If you go back to the past, you could run into dinosaurs or be caught in the middle of a medieval war; if you go to the future, you can find out people and cyborgs share the Earth!

Multiverse Theories

Have you heard about the story of Coraline? It is about a girl who reaches an alternative world that looks just like the ordinary one but hides a deep and dark secret. Even though this is just a film, the idea of many universes existing at the same time has intrigued scientists and inventors for a long time!

- This idea is called the multiverse, a mind-boggling concept that suggests our universe might be one of countless others existing side by side.
- Each universe could have its own unique laws of physics, different histories, and even different outcomes based on choices made differently than ours.
- Think about it—what if there was another universe where dinosaurs still roamed the Earth or where people had to learn how to fly instead of walk?
- Some theories of the multiverse suggest that these other universes are like bubbles floating in a vast sea, casually disconnected from one another.

- In some of these universes, the laws of nature might be entirely different.

- In one universe, gravity might be stronger, allowing giant creatures to float through the air, while in another, colors and light might behave in ways we've never imagined.

Why Do Scientists Think Multiverses Might Exist?

Scientists have developed different theories that could support the idea of a multiverse :

Quantum Physics

- This idea hints at the possibility of other universes existing where different choices have been made, or different events have occurred.

- Quantum physics is the study of very small particles, like atoms and subatomic particles.

- At this tiny scale, the behavior of particles is often strange and unpredictable.

- Quantum particles can exist in multiple states at once, a concept known as superposition. If we think about it, this means that anything is possible.

- One explanation of this behavior is that every time a decision is made or an event occurs, the universe splits into several different versions.

- Each version represents a different outcome.

- In simpler terms, imagine you are deciding whether to go to a park or stay home. In one universe, you choose to go to the park, while in another, you decide to stay at home.

- These different choices create different realities.
- According to this theory, there could be countless versions of ourselves making different choices in different universes.

Cosmic Background Radiation

- Another explanation for multiverses comes from cosmic background radiation.
- This radiation is a faint glow that fills the universe and can be detected from Earth.
- Scientists found this radiation when they were studying the Big Bang, which was the event that created our universe.
- The patterns and fluctuations in this radiation can tell scientists a lot about the early universe. These patterns may point to the existence of other universes.
- For example, if we observe anomalies in the cosmic background radiation, it might indicate interactions with other universes.
- This could happen if our universe is part of a larger system. If other universes exist and have their own forms of radiation, they could influence the cosmic background we see today.
- Thus, cosmic background radiation not only helps us understand our own universe but may also reveal clues about the possibility of multiverses.

Fun Facts about Multiverses

In some superhero films and TV shows for children, we see different universes where various superheroes interact with one another or face challenges in unique ways.

- The Amazing World of Gumball explores various dimensions where characters face alternate versions of themselves.
- When characters meet alternate versions of themselves, it often leads to moments of reflection. They may question who they are and invite you to engage with your own identities.
- In some chapters, characters might meet a version of themselves that makes different choices, leading to vastly different lives. What would you do if you could jump from one universe to another and make different decisions in each of them?

What If...?

If you could step through a magical door and enter an alternate universe where everything is different—like floating islands, talking animals, or even a world where time flows backward—what exciting adventures would you go on, and what kind of unique friends would you meet along the way?

The Phantom Time Hypothesis

This theory suggests that parts of history may have been fabricated or altered.

- It argues that there are inconsistencies in historical records, buildings, and calendars could imply that specific eras may not have occurred at all.
- The Phantom Time Hypothesis was introduced by a German historian named Herbert Illig in the 1990s.
- Illig proposed the idea that a period of time, specifically the early Middle Ages, was added to the calendar.

Did Medieval Knights and Princess Truly Exist?

- This means that the years we believe to be part of history from around the 8th to the 11th centuries did not actually happen as we think they did.
- Illig posits that certain emperors, particularly those in the Holy Roman Empire, may not have existed at all.
- He suggests that historical scholars created false documentation to support this altered timeline.

His Arguments

- Herbert Illig's bases his theory on careful analysis of historical records. He looked at various documents, artifacts, and the way history was recorded, and he noticed inconsistencies.
- He pointed out that there are very few trusted records from those centuries. Illig believed this gap in records suggested that the time was fabricated or misunderstood, rather than simply lost.
- He explained that the Gregorian calendar, which we use today, was not adopted until much later. Before it, the Julian calendar was prevalent.
- The shift to the Gregorian calendar changed the way time was measured and opened the debate: How do we know that the historical dates are correct when the systems used were different across cultures?
- Illig also pointed out the limited archaeological evidence for certain events and figures that we believe existed in those centuries.
- Many medieval texts, which describe significant events, appear to have been written much later than they supposedly occurred.

- For instance, accounts of major battles or kings who were thought to have lived during the early Middle Ages are often based on later writings. Did that really happen, or was it just a creative invention of some historian?

- Illig also examined the styles of art and noted a notable lack of development or innovation during what is considered the early Middle Ages.

What Do Illig's Colleagues Have to Say?

- Many historians just don't agree with Illig: They claim that his theory doesn't have enough evidence.

- They also argue that it is impossible for a squad of trickster scholars to have created such a large-scale deception.

- They highlight numerous artifacts and writings from different cultures that document events that happened during the Middle Ages.

- Could a group of archeologists place the ruins of the castles and bury the remains of the epic battles?

- History scientists don't believe that! Conversely, they are certain that the archeological digs across various regions are proof of the timeline as we know it.

- The other argument against the Phantom Time Theory is the purpose. Why would historians want to invent a whole era?

What If...?

What if the Phantom Time Theory is true, and we suddenly discovered that prehistory was entirely fabricated: People never lived in caves and didn't hunt mammoths because those never existed as well—how would that change our understanding of history and ? the evolution of societies?

Chapter 2

Epic Journeys Through Shadows and Secrets

Are you ready to embark on epic journeys to explore the unknown and uncover age-old mysteries? In a world filled with shadows and secrets, countless tales are waiting to be discovered!

From ancient artifacts that hold the key to lost civilizations to stories of vanished cities and mysterious regions shrouded in mystery, these journeys invite us into a realm where history, science, and imagination beautifully intertwine.

The Rosetta Stone's Role in Decoding Hieroglyphs

Have you ever heard about encrypted messages and unbroken codes? Imagine a giant rock covered in mysterious writing. That's the Rosetta Stone!

An Outstanding Discovery

- nearthed in 1799 by a French soldier during Napoleon's campaign in Egypt.
- At first, Bonaparte's soldiers didn't know its true value.
- Later, French scholars realized this stone might hold the key to unraveling secrets lost for over a thousand years.
- This seemingly ordinary slab of stone became a powerful gateway to understanding a whole culture and its unique writing systems of the ancient Egyptians.

The Ancient Egyptians

- The ancient Egyptians are best known for building large pyramids, which served as tombs for their pharaohs.
- These structures were filled with treasures and artifacts meant to help the pharaohs in the afterlife.
- The construction of these pyramids required a lot of planning and organization.
- Besides great constructors, ancient Egyptians had many achievements: Writing was another crucial element of ancient Egyptian culture.
- The Egyptians developed hieroglyphs, a system of writing that combined pictures and symbols.
- This writing system was used for religious texts, government records, and monumental inscriptions.
- Scribes, who were trained to read and write, held important positions in society.

- They recorded everything from tax collections to royal decrees, playing a key role in the administration.
- The civilization was one of the greatest empires in Antiquity and later, they were conquered by the Greeks led by Alexander the Great.

A Secret in Three Languages

- The Rosetta Stone is like a giant, ancient puzzle with three sections of writing, each using a different script.
- The top part features Egyptian hieroglyphs, a beautiful script that looks like tiny pictures.
- Below that is Demotic, a simpler form used for everyday purposes, and at the bottom, Ancient Greek.
- Why were these scripts together? Well, back in 196 B.C.E., a group of Egyptian priests decided to honor their king, Ptolemy V, with a decree praising his efforts and victories.
- They wrote it using all three scripts to ensure everyone could read it, whether they spoke the language of the gods, the common tongue, or the Greek language of the ruling elite (Solly, 2022).

Breaking the Code

- Now, decoding these scripts wasn't easy. Imagine getting a big jigsaw puzzle, but you have no idea what the final picture should look like. That's what scholars faced. P
- People knew how to read Ancient Greek, so they thought, "Great, this will be easy!" But they were wrong.

- It took nearly two decades of hard work and clever thinking to crack the code completely.
- Enter Jean-François Champollion, a brilliant linguist from France.
- Champollion dedicated much of his life to solving this mystery. He studied lots of languages, including Coptic, which helped him understand Egyptian hieroglyphs.
- In the early 1820s, he made an incredible breakthrough. Champollion discovered that hieroglyphs weren't just symbols but also represented sounds—like letters in our alphabet.
- With this realization, he finally unlocked the door to reading hieroglyphs, opening up a treasure trove of knowledge about ancient Egypt.

Why Is All This Important?

Well, the Rosetta Stone wasn't just a cool archaeological find; it changed how we view history and language. The Rosetta Stone allowed us to read the Egyptian hieroglyphics!

- Before this stone was discovered, hieroglyphs were a total mystery. People had no idea what these intricate carvings on temples and tombs meant.
- Thanks to the Rosetta Stone, scholars began to understand an entire civilization, learning about their daily lives, beliefs, and achievements. It's like finding a missing piece of history's jigsaw puzzle!

But that's not all. The Rosetta Stone has had a lasting cultural impact too.

- It's so famous that people even use "Rosetta Stone" today as a phrase. When folks talk about finding a Rosetta Stone, they mean discovering something that helps explain or decode a big mystery.

- You may have even heard of language-learning software named after it, showing how this ancient artifact still influences modern life.

wasn't

What if the Rosetta Stone could talk? It would share the secrets of ancient Egypt, telling stories of pharaohs and hieroglyphs, helping us understand how people lived thousands of years ago! What would you like to know about the ancient Egyptians and Greeks?

Mysteries of the Lost City of Atlantis

You have probably heard about Aquaman and his underwater kingdom. This half-human, half-Atlantean superhero, battles to unite the underwater kingdom of Atlantis with the surface world while uncovering his true heritage and protecting both realms from danger. But, did Atlantis ever exist?

Plato's City

- The tale of Atlantis begins with the philosopher Plato around 360 B.C.

- Plato told a story of this lost city that sparked endless curiosity and wonder for centuries. He never said it was just an invention of his imagination.

- Plato was an ancient Greek philosopher who described the fascinating city in his dialogues "Timaeus" and "Critias."

- Atlantis was portrayed as a powerful island kingdom that mysteriously vanished beneath the ocean waves.
- For over two thousand years, people have been captivated by what remains one of history's most intriguing narratives.
- Plato described Atlantis as an advanced civilization existing around 9,600 B.C.
- According to him, it was rich, prosperous, and larger than Libya and Asia combined. Do you know where that place is?

Only a Metaphor?

- Some believe Plato used the story as a moral lesson on society's dangers when consumed by greed and hubris.
- Others argue that Atlantis was based on actual places or events, pointing to similarities with known historical occurrences like the eruption of Santorini or other ancient disasters.
- Various theories about Atlantis' precise location on sites in the Mediterranean Sea and others in Antarctica.
- Some researchers think it lay near the Azores in the Atlantic Ocean, while others speculate about a connection to Minoan Crete, whose collapse could mirror Atlantis' demise.

Searching Atlantis on the Radar

- Explorers and scientists are continually driven by the allure of uncovering Atlantis.
- Modern technology aids these efforts, with sonar and underwater drones mapping the ocean floor in search of clues.

- These technologies provide fresh hope that one day, we might discover remnants of this legendary city.

- Expeditions continued to scour the depths of Morocco and Spain, searching for evidence that could confirm Plato's tales even though there isn't evidence to show the city ever existed.

Imaginary Visits to Atlantis

Art, literature, and film have all drawn inspiration from this captivating myth of Atlantis. There are many popular films, comics, and books depicting Atlantis, in part as described by Plato, and in part as imagined by artists.

- ***Atlantis:*** Milo's Return: This direct-to-video sequel to "Atlantis: The Lost Empire" continues the adventures of Milo and his friends as they explore more about the mysteries of Atlantis.

- ***SpongeBob SquarePants:*** The beloved animated series often references Atlantis in various episodes, portraying it as a whimsical underwater city.

- ***Justice League: Throne of Atlantis:*** An animated film that explores Aquaman's origins and his connection to the underwater city of Atlantis.

- ***The Marvel Comics Universe:*** Atlantis appears in various comic series, including stories featuring characters like Namor the Sub-Mariner, who is the king of Atlantis.

Artists depict majestic cityscapes submerged underwater, while writers weave tales of discovery and loss that resonate with the allure of the lost city.

What If...?

What if Atlantis really existed, and we could dive deep into the ocean to explore its glowing streets and meet friendly mermaids? Imagine discovering ancient treasures and learning about the magical powers of the Atlanteans while making new underwater friends on our adventure!

The Enigmatic Bermuda Triangle

Can a plane simply vanish in the sky? Could a ship be swallowed by the stormy sea? Unfortunately, there are many stories of aircraft and vessels missing in a specific area in the Atlantic Ocean. It is called the Bermuda Triangle.

The Bermuda Triangle's Location

- The Bermuda Triangle is situated roughly between Miami, Bermuda, and Puerto Rico.
- Geographically, the Bermuda Triangle has unique maritime conditions.
- This area's boundaries, forming a rough triangle, include the warm Gulf Stream, which is known for its unpredictable weather patterns and fast-moving currents.
- In addition, the Sargasso Sea is located within the Bermuda Triangle. It has large quantities of floating seaweed and calm winds.
- Mariners have historically found these waters difficult to navigate due to their unique and sometimes perilous conditions, which can lead to disorientation and accidents at sea.
- This infamous region is notorious for its perplexing history filled with unexplained vanishings of diverse vessels.

Most Notable Incidents

Among the most notable incidents are the disappearances of Flight 19 and the USS Cyclops.

- Flight 19, a squadron of five TBM Avenger torpedo bombers, vanished during a routine training flight in December 1945.
- The flight's leader, Lieutenant Charles Taylor, reported that the planes were disoriented and running low on fuel.
- Despite the plane being led by an experienced pilot, the attempts to navigate back to base failed, and they vanished without a trace.
- In March 1918, the USS Cyclops, a massive navy cargo ship with over 300 men and thousands of tons of manganese ore aboard, disappeared without a trace.
- Despite being equipped with radio communication devices, no distress signals were sent.
- The Cyclops' disappearance remains one of the largest losses of life in U.S. Naval history not caused by combat, further intensifying the air of mystery surrounding the triangle.
- Many other planes and ships went missing in the Bermuda Triangle:
    - Flight 11 (1948): A Douglas DC-3 aircraft vanished while en route from Puerto Rico to Miami, with no distress signal or trace found.
    - The USS Scorpion (1968): A nuclear submarine that sank in the Atlantic Ocean, near the Bermuda Triangle, under mysterious circumstances, with all 99 crew members lost.

- The MV Joyita (1955): A merchant vessel that disappeared in the South Pacific near the Bermuda Triangle, was later found adrift with no crew onboard, and mysterious circumstances surrounding its disappearance.
- The Star Tiger (1948): A British South American Airways passenger plane.

Attempts to Explain the Mystery

Many theories have tried to explain what causes these mysterious disappearances. Nobody wants to fly or sail across the doomed triangle!

- Some scientists propose natural phenomena as plausible explanations.
- For instance, underwater methane hydrates, which can lead to sudden releases from the ocean floor, might cause changes in water density, resulting in ships suddenly losing buoyancy and sinking without a trace.
- Rogue waves, which are extremely large and unexpected ocean waves, have been known to occur in this region.
- These could reach towering heights, potentially overwhelming vessels or aircraft caught unaware.

Yet it's not just science that seeks to explain these mysteries: popular culture and folklore play substantial roles too.

- Unconventional theories have created stories of supernatural events, extraterrestrial involvement, and portals to other dimensions.

- According to the scientific agency of the United States government, the apparent high frequency of disappearances may simply reflect greater traffic in the area.
- This probably goes along with human errors, mechanical failure, and natural environmental factors.
- Statistics showcase that the number of incidents isn't significantly different from any other heavily traveled region of the world.
- Nevertheless, these rational explanations do little to quell the excitement and curiosity surrounding the Bermuda Triangle.

Excursions into this area continue to be pursued by adventurers, researchers, and curiosity seekers alike, all hoping to unlock the secrets behind its enduring mystique.

What If...?

What if we could uncover the secrets of the Bermuda Triangle and reveal the truth behind its mysteries? Imagine a team of brave explorers who use magic maps and special gadgets to dive into the ocean, discovering sunken ships, and ancient clues that explain why so many boats and planes have disappeared over the years! Would you dare to join the team?

Chapter 3

At the Speed of Light—Wandering the Cosmos

Whenever we stare at the starry night in the middle of the night, countless questions come to mind: How was the universe born? Is it still growing? Is there any type of life out there? Where do all those lights and colors come from?

Imagine the universe as an enormous sandbox, infinitely expansive, where every grain reveals stories from both the past and the future. That's what we are about to discover now!

The Expanding Universe

The universe is a vast and mysterious place, and one of the most intriguing discoveries about it is that it is constantly expanding.

- Imagine blowing up a balloon; as you blow more air into it, the surface stretches out in all directions.
- Similarly, our universe is stretching out, but on a much grander scale!
- This idea of an expanding universe gives us clues about how everything came to be and where it might be headed.

The Redshift Phenomenon

Why do we think galaxies are speeding away from us? The answer lies in something called the redshift phenomenon.

- When we look at light from faraway galaxies through telescopes, we see that their light has shifted towards the red part of the spectrum.
- This redshift means the light waves are stretched out, a sign that the galaxies are moving away.
- It is similar to hearing a siren sound change pitch as an ambulance drives past—a concept known as the Doppler effect.
- In the case of galaxies, the farther they are, the more their light is redshifted, confirming that the universe is indeed expanding continuously over time.

It All Started With a Boom... Or With a Bang!

But how do we know this expansion started in the first place? Here's where the cosmic microwave background radiation comes into play.

- This invisible radiation fills the universe and is evidence of the initial explosion and later stretching of space.

- Modern highly sensitive telescopes capture this radiation, and it tells us that once, a long time ago, everything was compact before expanding to the vastness we see today.

Speeding Up!

The universe's rate of expansion is actually speeding up which led the scientists to review some of the accepted theories.

- Not long ago, scientists believed that the universe would either keep expanding forever at a decreasing rate or eventually stop and reverse course due to gravity.

- New discoveries led them to admit that the universe will continue to expand indefinitely, leading to a future scenario often referred to as the "Big Freeze."

- The Big Freeze, the reversed Big Bang phenomenon, will be a scenario where galaxies drift further apart, and stars run out of fuel.

A Glimpse Into the Past

In earlier chapters, we discussed the hypothesis about traveling in time, remember? Well, telescopes that show us distant galaxies are another way to travel back in time. The light shown in the telescopes comes from a long time ago!

- Looking at the universe is like peering into the past, as light from distant stars and galaxies takes time to travel to Earth.

- When we observe celestial objects, we are not seeing them as they are now, but rather as they were when the light started its journey.

- For instance, light from the Sun takes about 8 minutes to reach us, so we see the Sun as it was 8 minutes ago.

- Similarly, light from the nearest star, Proxima Centauri, takes over four years to arrive, meaning we see it as it was four years in the past.

- For more distant galaxies, the light can take millions or even billions of years to reach us, providing a view of the universe at various stages of its history.

Thus, the farther we look into space, the further back in time we are observing.

The Hubble Space Telescope

Launched in 1990, Hubble orbits Earth and has provided breathtaking images and invaluable data about the cosmos.

- Hubble's mission includes observing distant galaxies, nebulae, and other astronomical phenomena, allowing scientists to study the universe's expansion, cosmic background radiation, and the formation of stars and galaxies.

- By examining the light from these distant objects, Hubble has helped establish that the universe is approximately 13.8 billion years old, a foundation for the Big Bang theory.

- The James Webb Space Telescope plans to look back in time to learn more about the universe's first moments, the creation of the first stars, and how galaxies began.

What If...?

What if we could watch the past of the universe like a movie, and by looking back at the Big Bang, we could discover hidden secrets about how stars and planets were born, and even learn why our own planet Earth was formed?

Dark Matter and Dark Energy—May the Force Be With You!

Dark matter and dark energy are mysterious forces in the universe, much like the Force in Star Wars! Together, dark matter and dark energy make our universe an exciting and mysterious place, just waiting for young Jedi explorers like you to discover more!

Dark Matter

- Just as the Jedi use the Force to move objects and have special powers, dark matter helps hold galaxies together and keeps them spinning.
- It is invisible, but we know it is there because of the way stars dance around it.
- Imagine space as a grand stage where galaxies are the main performers, spiraling in graceful cosmic ballet: Something unseen is pulling their strings—dark matter.
- It is like an invisible glue keeping everything together!
- How do we know it exists if we can't see it?
- Scientists can infer its presence through the gravitational effects on stars and galaxies.
- For instance, when observing galaxy rotation speeds, astronomers find them moving in ways that suggest much more mass exists than what we can observe with telescopes.

- The hunt for dark matter involves elaborate experiments deep underground or in space stations just beyond Earth's atmosphere.

- One famous experiment, the Large Hadron Collider (LHC), smashes particles at incredible speeds to possibly create brief flashes of dark matter.

Dark Energy

Dark energy is like the Emperor, causing the universe to expand faster and faster, pushing galaxies away from each other, just like how the Sith want to control everything.

- While dark matter holds things together, dark energy seems to work in the opposite way, pushing the cosmos apart.

- It is believed to drive the accelerated expansion of the universe.

- Imagine blowing up a balloon; dark energy acts like the air within, causing the universe to expand faster over time.

- Astrophysicists, the scientists who study the universe, have observed that every galaxy is moving away from each other at increasing speeds. They know it is dark energy's powerful influence.

But what exactly is dark energy? That's still a big question mark.

- Diverse observations and theories try to explain what dark energy exactly is, but many questions remain unanswered:
    - Could it be a new kind of field or particle?
    - Or is it simply a property of space itself?

What If...?

What if dark matter is like the secret allies of the Resistance, holding everything together in the galaxy, while dark energy is like the First Order, constantly pushing and pulling to expand the universe even more, causing chaos and change in the cosmos?

Solar Storms and Their Impact on Earth

Imagine a storm on earth—Aasymphony of thunder and flashes of lightning, accompanied by a refreshing breeze, nourishes the blooms with gentle rainfall. T. However, other types of very different storms take place far away from us, though they still impact our lives: Solar Storms.

- They are intense bursts of energy and particles released from the sun.
- Solar storms are often a result of solar flares or coronal mass ejections (CMEs).
- Solar flares are sudden, intense bursts of radiation and energy from the Sun's surface.
- They occur when magnetic energy that has built up in the Sun's atmosphere is released, resulting in bright flashes of light across the electromagnetic spectrum, including visible light, radio waves, and X-rays.
- Coronal mass ejections (CMEs), on the other hand, are massive clouds of solar plasma and magnetic fields ejected from the Sun's corona into space.
- These storms can send millions of tons of charged particles into space.
- Despite these being huge phenomena, they can't be seen by the naked eye from the Earth.

- However, we can see some of the effects: They can create beautiful phenomena like the Northern and Southern Lights (auroras).

Bad (Sun) Weather

Solar storms can have many impacts on Earth, causing disruptions in communication and technology. They are a fascinating reminder of the powerful forces at work in our solar system!

- Solar flares and coronal mass ejections (CMEs) can disrupt communication and GPS systems.
- When these phenomena interact with Earth's magnetosphere, they create geomagnetic storms that can interfere with satellite operations.
- This interference affects GPS systems relied upon by various industries such as aviation, maritime navigation, and agriculture.
- Imagine airplanes unable to pinpoint their exact location or ships at sea navigating blindly—these are potential scenarios during severe solar events.
- High-altitude flights, especially those over polar regions, encounter increased levels of radiation during solar events.
- This heightened exposure poses health risks to both passengers and crew members.

Solar Storms Affect the Weather on Earth!

- Solar storms also influence atmospheric conditions.
- As geomagnetic storms interact with Earth's atmosphere, they can alter weather patterns and ionospheric activity.

- The ionosphere, a region filled with charged particles, plays a crucial role in radio wave propagation.
- Disturbances here can affect radio communications, impacting areas like emergency response coordination and ham radio transmissions.

What If...?

What if solar storms were like the Sun throwing an epic cosmic party, sending colorful bursts of energy that light up the night sky with dazzling auroras? Imagine you could be a young space explorer, watching for these spectacular displays and discovering how our amazing universe works—are you ready to explore and learn more about the wonders above?

Chapter 4

Sounds to My Ears—Magical Melodies

Did you know that there is a way to make the time go slower or faster? With music! Studies have shown that when listening to fast-paced music, people tend to perceive time passing more quickly, while slower music can make time feel like it is dragging.

Join us on an exploration of how music has evolved over centuries, from ancient instruments to modern compositions, revealing its remarkable ability to shape our emotions and thoughts.

The Invention of the Piano

Even if you know very little about music, you would still be able to recognize the sound of a piano. Imagine a world without the piano, the centerpiece of countless living rooms and concert halls.

- The piano's journey began with Bartolomeo Cristofori, an Italian inventor who had a knack for creating musical instruments.
- Around 1700, Cristofori changed music forever by introducing this magnificent instrument.
- Before the piano, musicians had to stick with the soft sounds of a clavichord or the louder but less expressive notes of a harpsichord.
- The magic of the piano lies in its ability to be both gentle and powerful, thanks to its unique hammer mechanism.
- With this feature, musicians could play notes that were whisper-soft or thunderously loud.

The Emotions Creator

Cristofori's invention was truly revolutionary because it opened up new possibilities for musical expression.

- This was highlighted in his early creation, called the *clavicembalo col piano e forte*, which means harpsichord with soft and loud.
- The piano was not just an invention; it was a game-changer.
- It allowed composers to experiment with volume and emotion in ways they never could before: Imagine a piece where a single note can evoke a tear or a smile, depending on how it's played.

An Evolving Design

As years passed, the piano evolved dramatically from Cristofori's original design.

- First, they were known as fortepianos, which couldn't match today's instruments in terms of strength and precision.

- During the Industrial Revolution (1770s-1850), the piano underwent significant changes because the new technologies allowed the use of different materials.

- Improved manufacturing techniques made pianos more robust and more affordable, allowing them to be produced and sold en masse.

- By the late 19th century, having a piano was almost as common as having a front door!

- Since people had a piano at home, it was easier for everybody to learn music, and many people became musicians.

This evolution paved the way for composers to write more complex and rich compositions.

The Age of Composers

The end of the 18th century, when the Industrial Revolution began, and the 19th century became the age of the composers. The most popular composers of classical music created their masterpieces in this period using a piano.

- Beethoven composed the Moonlight Sonata' and Fur Elise, two iconic pieces that still captivate us today.

- Chopin's delicate waltzes, *Grande Valse Brillante* and the *Minute Waltz* were composed especially for piano.

- Liszt's challenging études, such as the *Hungarian Rhapsody* and *Dreams of Love.*

- Schumann's evocative songs, the *Scenes From Childhood* and *Carnaval* were also composed on a piano.

The Piano in Modern Music

Beyond classical music, the piano's versatility shines brightly across various genres.

- Jazz musicians, for instance, have embraced the piano as a core element since the genre's early days.
- Legends like Duke Ellington and Thelonious Monk used the piano to explore complex harmonies and rhythms.
- In rock'n roll and pop music, artists like Elton John and Billy Joel brought the piano to center stage.
- The evolution of keyboards from the traditional piano has transformed the landscape of modern music. The British band Coldplay uses the keyboards and synthesizers, which have become a hallmark of their sound.

What If...?

What if the piano could talk and share the stories of all the songs it has played? Would you rather chat with Beethoven's piano or Elton John's? Or would you prefer to interview Coldplay's keyboards?

The Weirdest Musical Instruments in the World

The Theremin

Have you ever heard of a musical instrument that you don't need to touch to play? Meet the theremin! This curious contraption looks like something from a science fiction movie.

- It is made up of two metal antennas and some electronic components, but what makes it truly magical is how you play it.
- You create sound by disturbing electromagnetic fields only by moving your hands in the air around the antennas.
- This unique way of producing music results in eerie tones that have been used in spooky film scores and sci-fi television shows for decades.

The Hang Drum

- Now, let's tap into the world of percussion with a couple of instruments that take creativity to new heights.
- The Hang Drum is a UFO-shaped steel drum played with the hands, much like a bongo, but with one big difference: it sings.
- When struck, it produces rich, melodic harmonics that can mesmerize anyone who listens.
- It was invented in the early 2000s by Felix Rohner and Sabina Schärer in Switzerland.

The Waterphone

- This is an instrument that looks as unusual as its name suggests.
- The waterphone is an unconventional instrument invented by Richard Waters in the 1960s, known for its haunting and ethereal sound.
- Filled with water, it creates hauntingly beautiful, unsettling sounds when its metal rods are struck or bowed.

Australia's Didgeridoo

- This wind instrument is believed to be over a thousand years old, invented and played by Indigenous Australian people.
- The Australia's didgeridoo is made from hollowed-out eucalyptus tree trunks.
- Indigenous people used eucalyptus trees that were hollowed out by the termites.
- It produces deep, resonant drones that mimic the sounds of nature, creating a profound connection to the land and its people.

What If...?

Musical instruments are designed and crafted by skilled people called luthiers. They have a deep understanding of materials, acoustics, and construction techniques required to create instruments that produce beautiful sounds.

What if you could be a luthier for a day? Which sounds would you like to recreate? Would you like to create wind or string instruments? What type of materials would you like to try out?

Mysterious Effects of Music on the Brain

Magic happens when melodies float through the air, dancing into our ears and settling within our hearts. Listening to music is like discovering a fascinating secret about ourselves!

Music and Mood

Have you ever noticed how listening to your favorite song instantly brightens your mood or helps you relax after a long day?

- Music has a magical way of influencing our emotions.
- Studies have shown that listening to music can stimulate the release of dopamine, a neurotransmitter associated with pleasure and reward.
- This release of dopamine can create feelings of happiness and joy.
- It is no wonder why upbeat tunes are often used in music therapy to help people feel happier and reduce stress.

Brain Waves and Beats

Different types of music have unique effects on our brains, affecting how we think and focus.

- Classical music, for example, is known to produce a calming effect, slowing down brain waves and promoting relaxation.
- This is particularly useful during activities that require deep concentration, like reading or studying.
- On the other hand, fast-paced music, such as rock or electronic dance music, can quicken brain activity.
- It boosts alertness and hones focus, perfect for tasks where energy and attention are key.

Memory Magic: The Power of Jingles

Memories have a funny way of intertwining with music. Have you ever heard an old song and suddenly remembered a moment from years ago?

- This connection between music and memory can be incredibly powerful.

- Many advertisers use simple, catchy jingles for their products because they're easy to remember.
- These jingles stick in our minds, making it hard to forget the product associated with them.
- Our personal lives are filled with similar associations.
- The tunes our parents played when we were young, or the songs from summers past, often become tied to specific memories.

Music and Creativity

When it comes to creativity, music is like rocket fuel for the imagination.

- Engaging with music can stimulate the brain's creative centers, sparking new ideas and inspiring artistic endeavors.
- It is no surprise that many artists, writers, and inventors turn to music when they need to tap into their creative potential.
- When you are feeling stuck or uninspired, try putting on a piece that resonates with you—whether it's a lively jazz tune, an epic movie score, or a soulful ballad.
- This auditory backdrop can unlock imaginative possibilities and encourage creative expression across various fields.

Music can be a powerful tool for enhancing our mental processes and well-being. It offers us entertainment, emotional healing, and cognitive benefits.

What If...?

What if you could use music as a superpower, where each note you played could help you fly like a superhero, calm wild animals with a soothing melody, or even make flowers dance in the breeze, transforming the world around you into a magical place filled with color and joy just by the sounds you create?

Chapter 5

Coloring the World

Art is everywhere, not just in galleries or museums but in everyday life, where culture and creativity meet effortlessly. When we explore these colorful worlds, we start to see details we might have overlooked before—like little secrets waiting to be discovered.

Have you ever wondered why some paintings leave such a lasting impression, or why certain cultural artifacts are revered by entire communities? Let's find out!

The Enigma of Mona Lisa's Smile

The Mona Lisa is one of the most famous paintings in the world, and its charm lies not just in its visual beauty but in the mysteries that surround it.

- It was painted by Leonardo da Vinci, and it showcases his revolutionary techniques and attention to detail.

- One of the techniques he used is sfumato, which involves gently blending colors and softening edges to create an almost smoky effect.

- This technique gives the Mona Lisa her lifelike appearance, showcasing da Vinci's dedication as he worked on the painting for several years.

Imagine spending all that time perfecting something so that it's just right! That's the kind of dedication artists like Leonardo put into their work.

The Mysterious Smile

What intrigues many people about the Mona Lisa is her smile.

- It is not a big, obvious smile; instead, it's subtle and seems to change depending on where you stand or how you look at it.

- This has puzzled people for centuries and sparked numerous interpretations about what emotions are being expressed. Is she happy? Sad? Thoughtful?

- The ambiguity invites viewers to think critically and form their own connections with the painting.

- It is like a riddle that doesn't have one correct answer, making each person's experience personal and unique.

Mona Lisa's Home

Mona Lisa traveled a lot before reaching her present-day home in France!

- Today, the Mona Lisa resides in the Louvre Museum in Paris, one of the most visited places in the world.

- Leonardo Da Vinci first took it to France, traveling on the back of a mule. He only had three of his paintings with him, one of which was Mona Lisa.

- In 1911, something unbelievable happened—the Mona Lisa was stolen!

- For two whole years, she was missing, which shocked people everywhere and added to the painting's already mysterious reputation.

- During World War II, it was taken out of the Louvre and hidden in a castle so the Nazis couldn't steal it.

The famous painting was restored to the Louvre after the end of the war and now stands in the largest room.

Other Intriguing Details

Did you know the painting is surprisingly small? It measures about 30 x 21 inches, which might seem tiny compared to its enormous reputation.

- There are various hidden details within the painting itself: Leonardo included little elements throughout the artwork, some visible only through careful examination or modern technology.

- The painting includes a detailed landscape with winding paths, a bridge, mountains, and a distant village.

- The woman is dressed in a dark. Lower-class women used dark clothes in the early 16th century. By then, only members of the nobility were portrayed... Who was this woman?

These details contribute to discussions about the painting and Leonardo's life.

What If...?

What if Leonardo da Vinci was not just painting a beautiful woman in the Mona Lisa, but secretly telling the story about an unknown important woman hiding her power with every smile and secret glance?

Round-Headed People in Algeria

In the vast and varied landscape of Algeria, we discover a fascinating artwork known as the round-headed people.

- This term may sound whimsical at first, but it actually refers to a specific cranial shape that has captured the attention of anthropologists and historians alike.
- The round-headed people are depicted in cave paintings in Algeria, in North Africa.
- Anthropologists believe the Capsians painted them, and they are associated with the Capsian culture that existed in North Africa during the Mesolithic period (approximately 10,000 to 5,000 years ago).

How do they get there?

- Even in the Age of Cavemen, people moved from one territory to another, seeking better lands and resources.
- Throughout time, trade routes crisscrossed the region, bringing together diverse groups and enhancing genetic diversity. Imagine caravans laden with goods, slowly making their way across deserts and mountains, linking distant lands and peoples.

- These movements weren't just about economics; they were cultural exchanges that left lasting imprints on the populations who lived along these routes.

What the Paintings Tell Us About the Capsians

These paintings bring us information about what these people were like, or at least, how they saw themselves and how they wanted to be remembered. The depictions also tell us about their lives and ventures.

- The Capsians are often characterized by their round skull shapes, which are distinctively part of the broader study of human physical diversity.

- The Capsians are known for their advanced tools and artifacts, including finely crafted stone tools, which indicate complex behaviors and technologies.

- The rock paintings and engravings found in various caves and rock shelters in Algeria, depicted animals, human figures, and various symbols.

- They were primarily hunter-gatherers and later began to adopt some agricultural practices.

- The Capsians are considered ancestors of modern populations in North Africa, and their influence can still be seen in the region's genetic and cultural makeup.

What If...?

What if you were a young anthropologist studying the round-headed people in Algeria, and you could travel back in time to their world—what exciting things would you want to know, like what games they played, what stories they told around the fire, or how they used their amazing tools to hunt and gather food in a land filled with adventure and mystery?

The Cultural Significance of Masks in Different Societies

We all love costume parties. Mascarades were popular celebrations in the past, too because people just love to play being someone else. But masks are much more than a simple entertainment accessory.

Masks are fascinating! They come in all shapes, sizes, and colors, and play important roles in many cultures around the world.

Art and Rituals

Masks are not just for fun; they have deep meanings too.

- In many rituals, masks represent spirits or gods. Wearing these masks is a way to connect with the spiritual world and honor ancient beliefs.

- Many African cultures, such as the Dogon of Mali and the Akan of Ghana, use masks in ceremonies and festivals to represent ancestors, spirits, or animals, often during dances that convey stories and cultural values.

- In Burkina Faso, the Bwa people create masks with stunning feathered headdresses. These masks are used in agricultural rituals and to mark the changing seasons.

- Balinese Kecak Dance In Indonesia, the Kecak dance features performers wearing masks that represent characters from the Ramayana epic.

- Greek Theater Masks: Ancient Greek theater utilized masks to represent different characters and emotions, allowing one actor to portray multiple roles. The oversized features of the masks enhanced the visibility of the audience.

- These ceremonies often include music and dance, creating a powerful and respectful atmosphere.

Masks and Identity

Masks serve as expressions of identity within societies. They can symbolize heritage, status, or specific cultural symbols.

- For example, in the Japanese Noh theater, the actors wear masks to portray different characters and emotions.

- Carnival of Trinidad and Tobago: During Carnival, participants wear vibrant masks and costumes to celebrate cultural heritage and freedom, often depicting historical figures, folklore, and social commentary through artistic expression.

- In Mexico, during the Day of the Dead celebration, people wear calavera masks, which depict skulls. Although they might seem spooky, these masks are actually cheerful, and colorful, and used to honor loved ones who have passed away.

- Indian culture also has its share of intriguing masks. In places like Kerala, there are traditional dances called Kathakali, where performers wear elaborate masks and costumes.

- The Native American cultures of North America have their own unique masks too.

- The Inuit people create masks as part of their storytelling traditions, sharing tales about animals, nature, and everyday life.

So, next time you see a mask, think about the stories it might tell. Why was it made? Who wore it? What does it symbolize?

What If...?

What if you could explore a magical world where every mask you found told a unique story—what would they reveal about the people who wore them, their dreams, and their adventures, and how could wearing one of those masks help you discover your own hidden talents and feel like a hero in your own story?

Stonehenge

Have you ever heard about gigantic stone monuments? They are called megaliths:

- The term "megalith" comes from the Greek words "mega," meaning large, and "lithos," meaning stone.
- These structures are often associated with prehistoric cultures, and they serve various purposes, including burial, ceremonial, or astronomical functions.
- One of the most famous groups of megalithic structures is found in the British Isles.

Stonehenge is one of the most famous prehistoric monuments in the world.

- It is located in Wiltshire, England, and it dates back to around 3000 BC to 2000 BC.
- This ancient site consists of a ring of standing stones that are set in the ground. There are massive stones, some weighing as much as 25 tons, arranged in a circular pattern.
- The stones in Stonehenge are a mix of local limestone and bluestone, which is believed to have been transported from Wales.

The Mystery of Its Construction

- It is thought that the construction of the monument involved a large workforce.
- People likely used simple tools and techniques to transport and carve the stones. They may have constructed sledges and used ropes to pull the stones over land.
- Waterways and logs might have assisted in moving the stones as well. The stones were then erected in position using ramps and wooden frameworks.
- Archaeologists have discovered burial mounds and other archaeological features in the surrounding area.
- Some of the nearby mounds contain skeletons, which points to the belief that this was a sacred area for burial and remembrance.

What Did They Build the Monument for?

The purpose of Stonehenge has been a mystery for many years.

- Some theories suggest that it was a place of worship, while others believe it had astronomical significance.
- The alignment of the stones indicates that the monument may have been used to mark the movements of the sun and the seasons.
- For example, during the summer solstice, the sun rises directly over the Heel Stone, which is one of the stones at the entrance of the site.
- This event attracts many visitors each year who come to witness the sunrise and celebrate the longest day of the year.

Other Stone Monuments

Stonehenge is not an isolated phenomenon. We can find similar megalithic structures across Europe and beyond.

- Italy
    - The Nuraghe are ancient stone towers found primarily on the island of Sardinia.
    - These unique structures, which date back to the Bronze Age, consist of large stones skillfully stacked together without the use of mortar.
    - Many Nuraghe served as fortifications, providing a defensive advantage to the communities that built them.
- Malta
    - The Megalithic Temples of Malta are some of the oldest free-standing structures in the world, with origins dating back to around 3600 BC.
    - These temples were made from large limestone blocks and were likely used for religious rituals or ceremonies.
    - The most famous of these temples is the Ħaġar Qim, which features intricate carvings and a layout that reflects the spiritual beliefs of the people of that time.
- South Asia
    - The megaliths of the Khasi tribe in Meghalaya, India, showcase a different style and purpose.

    - These structures often feature upright stones arranged in groups, sometimes marking burial sites or serving as memorials to the dead.
    - The Khasi people have a deep spiritual connection to these stones, believing they hold the spirits of their ancestors.
- Central America
    - The ancient city of Tikal, is located in present-day Guatemala.
    - Tikal features massive pyramids constructed from limestone blocks, and it was one of the major cities of the Maya civilization.
    - The largest pyramid, known as Temple I, towers over the surrounding jungle, reaching a height of about 70 meters.

What If...?

What if, in a world like the one in the *Ice Age*, a group of friendly prehistoric animals discovers an ancient megalithic structure similar to Stonehenge? What if the stones were put there by a race of giant creatures who once roamed the earth, or perhaps they served as a magical portal to other worlds?

The Inuit Tupilak

Tupilaks are traditional figures from Inuit culture, particularly associated with Greenland.

The Inuits

- The Inuit are a group of indigenous peoples primarily found in the Arctic regions of Canada, Greenland, and Alaska.

- The Inuit speak different dialects of the Inuktitut language, which belongs to the Eskimo-Aleut language family.

- The Inuit were semi-nomadic hunters and gatherers, relying on hunting marine mammals such as seals, whales, and walrus, as well as fishing and gathering wild plants during the short summer months.

Inuits Crafts

- Tupilaks are often crafted from materials like bone, ivory, wood, or stone and are typically designed to represent figures such as animals, spirits, or mythical beings.

- Tupilaks hold cultural significance and were traditionally believed to possess magical powers, often used in rituals or as talismans.

- They symbolize a connection to the spiritual world and the natural environment in Inuit culture.

The Creation Process

Creating a Tupilak involves several steps that highlight the deep spiritual beliefs of the Inuit people.

- First, the creator would select materials that resonate with their intentions. For example, they might choose a specific type of bone if they want to represent strength or resilience.

- After selecting the materials, the creator would carve or shape the Tupilak into a figure to invoke wisdom or courage.

- Once the figure is completed, the creator often takes time to imbue the Tupilak with meaning through rituals.

- These rituals can include words of power, prayers, or songs that are significant in Inuit culture.

This process is vital as it connects the spirit of the Tupilak to the creator, reinforcing the bond between them.

The Role of Tupilaks

Tupilaks serve various purposes within Inuit society.

- Protection
    - They can guard the creator and their family from harm.
    - When someone believed they were in danger or needed extra support, they might call upon their Tupilak for assistance.
- Identity
    - Additionally, Tupilaks can act as representations of personal qualities.
    - For instance, an individual may create a Tupilak that symbolizes their desire for bravery.
- Artistic Expression
    - Tupilaks often feature exaggerated features and abstract shapes. T
    - Some artists choose to include animals in their designs. For example, a Tupilak might have the characteristics of a polar bear to symbolize power and tenacity.

What If…?

What if, in a magical snow-covered land, you and your friends discover ancient Inuit tupilaks hidden in a glimmering ice cave? Each of you is invited to create your own special tupilak. What would your tupilak look like? Would it be a swift, playful spirit like a seal, a wise and gentle guardian like a polar bear, or perhaps a brave adventurer like a great eagle?

Chapter 6

Sports Euphoria

Run as if you had wings; Throw the ball as if you were Zeus throwing lighting; Swim as if you were chasing Aquaman... Sports euphoria goes much further than competition: It makes us feel stronger and powerful.

This chapter shares the tales of remarkable achievements and astonishing dedication carving out a path that leads us to view sports as more than games; they become chronicles of the human spirit!

Origin of the Olympic Games

Any athlete dreams of having a gold medal and climbing to the first place on the podium, but that wasn't the original award. Did you know?

- Over 2,700 years ago, in the heart of ancient Greece, a remarkable tradition took shape—the Olympic Games.
- These games were not merely athletic contests; they were rooted in religious festivals, celebrating both human spirit and divine reverence.
- Athletes from various Greek city-states gathered in Olympia to compete, aspiring for honor and glory.
- It was believed that athletes honored and appeased the gods who watched over these festivities.

Which Were the Original Olympic Sports?

- Initially, the Olympics featured only one event—a simple foot race called "stade," which was approximately 192 meters long.
- Over time, the Olympic Games expanded to include a variety of competitions.
- The pentathlon was a significant event that included five disciplines: running, long jump, discus throw, javelin throw, and wrestling.
- Boxing was also popular, with fighters using leather straps on their hands for protection. Matches continued until one fighter gave up or was knocked out.
- Pankration was a brutal sport that combined elements of boxing and wrestling, with few rules. Fighters could use any technique, including strikes and grappling, making it one of the most challenging events.
- Chariot racing was an equestrian event in which teams of horses pulled a chariot driven by an athlete. This event was particularly

dangerous and showcased the skill of both the driver and the horses.

Where Were the Girls?

For many centuries, women were barred from participating in the Olympic Games.

- The inclusion of women in the Olympic Games in 1900 marked a pivotal shift towards inclusivity and equality in the realm of sports.
- Female basketball and volleyball were accepted as Olympic sports in 1976.
- Women's soccer was first included in the Olympic Games during the 1996 event held in Atlanta.
- Female boxing was admitted to the Olympic Games in London in 2012.

It has taken a lot of time to give girls the same space as boys to compete in the Olympic Games!

The Sacred Fire of the Olympic Torch

- An integral part of the Olympics, both ancient and modern, is the Olympic torch relay.
- The torch represents the ideals of peace and unity.
- The journey of the flame, carried by numerous torchbearers, represents the bond that connects nations and people worldwide.
- When the torch reaches its final destination and lights the Olympic cauldron, it signifies the dawn of a new era of

international camaraderie, where athletes unite under the shared banner of friendship and competition.

What If...?

What if, in a magical world where all the great athletes of history come together, you were invited to participate in a brand-new Olympic Games? Who would you cheer for? Or would you be one of the athletes engaged in the competitions?

From China to the World—Soccer Rules' Evolution

Soccer, a sport adored worldwide today, has roots that stretch back thousands of years.

- Imagine ancient China around 400 BC, where a game called "Cuju" was played.
- Players would kick a leather ball through an opening into a net using a mix of skill and teamwork.
- This wasn't just a game; it was part of military training and cultural expression.
- Cuju shows us how different civilizations have contributed to what we know now as soccer.

The British in China learned about Cuju and brought it to their country.

Setting the Rules

- In the 19th century in England, soccer had many players and enthusiasts. They struggled because they didn't follow a unified set of rules emerged.

- The matches were played with very local and varied guidelines, leading to chaotic scenes on the field.
- The creation of the Cambridge Rules in 1848 by students at Cambridge University established key principles like not using hands, setting boundaries on the field, and defining offside positions.
- They laid the groundwork for what would become modern soccer.

Slowly, soccer turned into a professional sport, and the leagues became important championships.

Heated Controversies

Of course, one of the most debated aspects of soccer is the offside rule.

- This rule, designed to prevent players from lingering near the opponent's goal for easy scores, has seen numerous revisions over time.
- At first, it was vague, but then, the rule evolved to specify lines and player positions, adding strategy and complexity to the game.
- It can help the defense or the attack, depending on how it is used.

If a player scores, the goal can be claimed invalid if the referee detects they were offside. The whole course of a match can shift due to an offside position!

A New Challenge—The VAR

As technology advanced, so did soccer. The introduction of VAR - Video Assistant Referee - represents a significant technological leap.

- VAR allows referees to review decisions using video footage, ensuring accuracy in critical match moments such as goals, penalties, or red cards.
- While it aims to improve fairness, VAR has sparked debates among fans and experts about its impact on the game's flow.
- Some claim it supports transparency, while others affirm it disrupts the spirit of soccer.

The time we have to wait while the VAR checks a goal adds suspense and excitement to soccer matches!

What If...?

What if, during a big soccer tournament, the players had the chance to change some rules on the spot? Suddenly, they can use the VAR not just to review plays but to create unique rules for their game! What if they could challenge any referee's call using video clips of their best moves?

Native American's Sports

Indigenous sports have long embodied the community values and cultural richness of Native American tribes as symbols of tradition and identity.

Lacross

Let's first take a closer look at the sport!

- Players use a long-handled stick called a crosse, which has a netted pouch at the end used to catch, carry, and pass the ball.
- Protective gear, such as helmets, gloves, and pads, is also worn to prevent injuries.
- The objective of the game is to score goals by shooting a small rubber ball into the opposing team's goal. The team with the most goals at the end of the game wins.
- Lacrosse can be played on a grass or turf field. Teams typically consist of ten players in field lacrosse (three attackmen, three midfielders, three defenders, and one goalie) or six players in box lacrosse (five players and one goalie), which is played indoors.
- The game is divided into quarters, and players use their sticks to pass and catch the ball while navigating around opponents.
- Players can run with the ball, pass it to teammates, or shoot at the goal.

In the past, American Native people played it very differently...

- Lacrosse was more than just a sport; it was a deeply spiritual and culturally significant activity.
- Originating as early as 1100 AD, lacrosse was played by many tribes, notably the Haudenosaunee (Iroquois), across what is now the northeastern United States and parts of Canada.
- The game often involved hundreds of players and could stretch over miles and endure for days.

- Far from being just about scoring goals, these games were a conduit for tribal diplomacy, sometimes settling disputes or preparing young warriors for battle.

Shinny

Shinny has its roots in Canada and has been played for generations, often considered a precursor to organized ice hockey.

- Shinny was incredibly popular across various regions, including the plains and coastal areas, showcasing the universality of its appeal.
- The game involved two teams, the number of players varying greatly, often ranging from a few to several dozen.
- The objective was to hit a ball or a small object into the opposing team's goal, using sticks crafted from materials like wood or animal bones.
- Matches could take place on natural surfaces such as frozen lakes, open fields, or cleared areas in the woods. The playing field was often chosen for its accessibility and safety.
- While there were some common rules, the specific regulations could vary by tribe and region.
- Players typically used straightforward wooden sticks, and in some variations, they may have used deer hide for the ball or other materials available to them.

It was traditionally played by women, although men participated as well. Young warriors develop skills necessary for hunting and combat, such as hand-eye coordination, agility, and teamwork.

What If...?

What if, while exploring a vibrant forest, you stumbled upon a magical gathering of Native American children who invite you to join in their traditional sports? As you play games like shiny and lacrosse, you can discover the stories behind each sport and learn about the values of teamwork, respect, and community.

Chapter 7

All Aboard! Geographical Marvels

Exploring the geographical marvels of our planet is like embarking on a grand adventure without leaving your seat. Imagine diving deep beneath the surface into mysterious caves, or standing in awe before towering mountains that reach for the sky.

Nature has plenty of stories to tell, from the colossal peaks formed by the constant dance of Earth's tectonic plates to the whimsical weather phenomena that have puzzled and amazed us for ages. Each natural wonder gives us a glimpse into the powerful forces at work on our planet, inviting us to discover a world full of surprises and lessons about our environment.

The Formation of Mount Everest

You probably already know that Mount Everest is called the roof of the world because it is the highest point on Earth. Nonetheless, there is much more to learn! How did such a giant come to be? It is all thanks to something called plate tectonics.

Tectonic Plates Are the Ones to Blame

- Our planet's surface is like a massive jigsaw puzzle made up of huge pieces called tectonic plates.
- These plates are always in motion, even though we don't feel it. Sometimes, they crash into each other or pull apart.
- When these plates push against one another with great force, they can crumple and fold, creating towering mountains in their wake.

That's precisely how Mount Everest was formed.

The Birth of Mount Everest

- Millions of years ago, the Indian Plate collided with the Eurasian Plate.
- Imagine two giant clouds colliding and creating a big heap in the middle—only much slower and with enormous rocks!
- This incredible process doesn't stop.
- Mount Everest is still growing, rising by about 0.04 inches each year as these plates continue to press against each other.

Mountains' Voices

Some mountains 'talk' by creating earthquakes.

- The Everest region has experienced several major earthquakes, with the most notable being the 2015 Gorkha earthquake in Nepal, which had a magnitude of 7.8.

- This earthquake caused extensive damage and loss of life, impacting areas including the Everest region.
- Earthquakes can also trigger avalanches on Everest, as seen in 2015 when a massive avalanche swept through Base Camp.

The Mount of the Many Names

- It was during the Great Trigonometric Survey in the 19th century that people realized just how special this peak was.
- Back then, Everest was simply known as "Peak XV." Imagine calling the tallest mountain you could ever climb just by a number!
- Later, it was named after Sir George Everest, a British surveyor who helped map India.

Around the world, Mount Everest is known by different names.

- In Nepal, it is called "Sagarmatha," which means "Forehead of the Sky." Doesn't that sound grand and majestic?
- Meanwhile, in Tibet, it's known as "Chomolungma" or "Qomolangma," meaning "Goddess Mother of the World."

Each of these names adds another layer to Everest's story, showing how different cultures view and honor this remarkable giant differently.

What If...?

What if you could give Mount Everest a new name that reflects its greatness, like "Friendship Peak"? Would you dare climb it to the top? What adventures would you encounter along the way?

Bizarre Weather Phenomena Around the World

Weather events can sometimes feel like they're straight out of a storybook.. Imagine electric orbs lighting up the sky during a storm or fish raining down from above. Welcome to the fascinating world of meteorology!

Ball Lightning

Picture this: you're watching a thunderstorm when suddenly, glowing balls of light appear, floating in mid-air.

- This is ball lightning, a rare and mysterious weather phenomenon that has intrigued people for centuries.
- It occurs during thunderstorms and can last from a few seconds to several minutes.
- Despite its striking appearance, scientists still don't fully understand what causes ball lightning, which is why it continues to capture the imagination of those who witness it.

Some believe it's simply another form of lightning, while others think it might be something entirely different.

Raining Fish

- This event sounds too unbelievable to be true. However, it does happen!
- During severe storms, waterspouts—essentially tornadoes over water—can lift small sea creatures into the air.
- When the storm calms, these creatures fall back to the ground, often in areas far from their original location.
- These events have been reported all over the world:

    - Iguazu Falls, Argentina: There have been accounts of fish raining down in this area, likely carried by storms and winds over the falls.

    - Simington, Arkansas, USA: In the late 1800s, residents reported fish falling from the sky during rainstorms, leading to a local legend.

    - Lajes, Azores, Portugal: In this location, residents have observed fish falling from the sky, particularly small sardines, during certain stormy weather conditions.

    - Bangladesh: There have been reports of fish raining down during intense monsoon storms, often linked to strong winds and waterspouts.

Upside-down Waterfalls

Yes, it is what you believe: Water that falls up instead of down as it should! Where did gravity go?

- In certain parts of the world, where winds are incredibly strong, water flowing over cliffs can be pushed upwards, creating the illusion of a waterfall defying gravity.

- Imagine standing on a cliffside, expecting to see water cascading down, only to find it shooting back up into the air instead.

- Some of the most impressive and popular backward waterfalls are:

    - Cascata do Caracol (Brazil): This waterfall in southern Brazil has been reported to occasionally flow backward during strong winds, creating a stunning visual effect.

- Victoria Falls (Zambia/Zimbabwe): During high winds, sections of this massive waterfall can appear to blow water upwards, creating the illusion of an upside-down waterfall.
- Glen Canyon (U.S.): In certain windy conditions, water from falls in the Glen Canyon can also appear to flow backward, especially during storms.
- Angel Falls (Venezuela): While not commonly observed, there are occasional reports of the wind creating an upward effect on the water flow of this renowned waterfall.

What If...?

What if you were a young meteorologist for a day and had the chance to explore unique meteorological phenomena like supercell thunderstorms, ball lightning, or waterspouts? How would you study and gather data about these events, and what experiments could you conduct to understand the science behind these phenomena?

Stalagmites and Caves

Beyond our imagination, nature has the power to create unbelievable labyrinths where we could easily run into a prehistoric creature or find a tunnel to the center of the Earth. Step into the enchanting world of caves!

- Stalagmites are born from the steady drip of mineral-rich water over countless years.
- As each drop lands on the cave floor, it leaves a tiny deposit of minerals, much like how you might build a sandcastle grain by grain.

- Over time, these deposits grow into stalagmites, rising majestically from the ground.
- Stalagmites can come in various shapes and sizes, including tall, thin structures or short, rounded formations.
- Their appearance is influenced by the flow rate of the water and the mineral content.
- Stalagmites are primarily found in limestone caves, where calcium carbonate is abundant.
- They are often accompanied by stalactites, which hang from the ceiling of the cave.

Secret Hideouts

Caves are bustling ecosystems thriving in unexpected ways.

- Without sunlight streaming in, creatures such as bats make these dark nooks their sanctuary.
- Caves also shelter a hidden world of microbes, tiny organisms that have adapted to life without daylight.
- These microbes perform extraordinary feats, sometimes producing antibiotics that we rely on to fight infections.
- Human activities like tourism alter the cave environment. Visitors often leave behind more than memories—skin cells, hair, and clothing fibers, all adding nutrients that disrupt natural conditions.
- Pollution, too, poses a risk, with harmful chemicals seeping into caves, imperiling these fragile habitats.

Famous Caves

Here are some famous caves around the world known for their impressive stalagmites:

- Waitomo Caves (New Zealand): Known for its glowworm displays, Waitomo Caves also features spectacular stalagmites and unique limestone formations.

- Postojna Cave (Slovenia): One of the most famous show caves in the world, Postojna Cave is adorned with beautiful stalagmites, along with an extensive cave system.

- Lascaux Caves (France) While primarily known for its prehistoric cave paintings, Lascaux also contains notable stalagmites within its corridors.

- Giant's Causeway (Northern Ireland): This UNESCO World Heritage site features basalt columns and embedded stalagmites, showcasing natural geological formations.

What If...?

What if you embarked on an exciting expedition to explore the center of the Earth? Imagine you run into glowing crystals, ancient fossils, and hidden underground rivers. What would be the greatest challenge? Which would be the most dreadful creature you could find?

Earthquakes and Tsunamis in the World

Have you ever felt the earth beneath your feet is moving? Sometimes, it can be just a sense, but on other occasions, it might be an earthquake.

- Earthquakes are sudden, shaking movements of the ground that occur due to shifts in the Earth's tectonic plates.

- These tectonic plates are large sections of the Earth's crust that float on the semi-fluid layer beneath them.
- When these plates move, they can get stuck against each other due to friction.
- The stress from the gathering movement becomes too great, and the plates release energy, resulting in an earthquake.
- The result can be felt as a series of vibrations that can last from a few seconds to several minutes.

The Overwhelming Sea

After an earthquake, especially in coastal areas, there is a risk of tsunamis.

- Tsunamis are large ocean waves typically triggered by underwater disturbances, such as earthquakes, volcanic eruptions, or landslides.
- When an earthquake occurs under the ocean, it can displace a massive volume of water, leading to the formation of these waves.
- Tsunamis can travel across entire ocean basins at speeds of up to 500 miles per hour, making them incredibly dangerous and often difficult to predict.
- Tsunami waves can surge onto land, sometimes reaching heights of over 100 feet, flooding areas far inland, and causing severe damage to infrastructure, homes, and ecosystems

Destructive Power

Earthquakes and tsunamis reveal the destructive power that nature can show.

- The 2011 earthquake in Japan was one of the strongest ever recorded, measuring 9.0 on the moment magnitude scale.
- It caused severe damage, leading to buildings collapsing and significant loss of life.
- A minor earthquake, such as one measuring 3.0 or below, might only be felt by those very close to the epicenter, the specific point where it happens. These don't cause any significant damage.
- In 2004, a huge tsunami occurred in the Indian Ocean and scourged the coasts.
- This disaster was caused by a massive undersea earthquake near the northern coast of Sumatra, Indonesia.
- The resulting tsunami struck coastal regions across several nations, including Thailand and Sri Lanka, causing immense destruction and loss of life.

What If...?

What if you were a junior reporter covering an earthquake and tsunami safety event in their community? How would you interview scientists about how earthquakes happen and the steps to stay safe? What important messages would you share about earthquake preparedness, evacuation plans, and building a tsunami safety kit?

Hurricanes and the Origin of Their Names

The Caribbean and the coasts near the Tropics have the best beaches and crystalline waters. Nonetheless, these paradises can easily turn into very dangerous places when the heat and the surface of the ocean create hurricanes.

- Hurricanes are powerful storms that can cause catastrophic damage when they make landfall.
- These storms originate over warm ocean waters, typically in tropical regions, where conditions are just right for them to develop.
- As the warm air rises from the ocean's surface, it creates low-pressure areas that can lead to the formation of clouds and storms.
- The energy from the ocean fuels these storms, causing them to grow stronger as they move over the water.

Everything Needs a Name!

Naming hurricanes isn't just to have a record; it reduces the confusion in the public, especially if two or more storms occur at the same time when the media and social nets inform about the weather.

- There is a system to name hurricanes. The World Meteorological Organization (WMO) maintains a list of names that are assigned to hurricanes.
- Each year, these lists are used on a rotating basis, which helps in creating a standardized way to refer to storms.
- Each hurricane season, which spans from June 1 to November 30 in the Atlantic Ocean, includes a predetermined list of names.

- These names are alphabetically arranged and alternate between male and female names.

- For instance, the first storm of the season may be named Alex, while the second could be named Bonnie.

Names That Won't Be Repeated

- Sometimes, names of particularly destructive hurricanes are retired to avoid confusion in the future.

- Hurricane Katrina wreaked havoc in 2005, and its name was never used again for future storms.

- Other names that were removed from the list:

    - Sandy (2012): Known as "Superstorm Sandy," this hurricane severely impacted the East Coast of the United States, causing widespread destruction, particularly in New Jersey and New York.

    - Ike (2008): This hurricane caused significant damage in Texas, particularly in Galveston, as well as in parts of the Caribbean. It resulted in dozens of fatalities and extensive destruction.

    - Rita (2005): Striking just weeks after Hurricane Katrina, Rita caused significant damage along the Texas and Louisiana coasts, leading to flooding and the evacuation of millions.

    - Harvey (2017): Known for causing unprecedented flooding in Houston and surrounding areas, Harvey resulted in significant loss of life and property damage.

The practice of retiring names honors the impact that these storms have had and helps to show respect for the communities affected by them.

What If...?

What if you could create a new system to name the hurricanes? What would be the criteria? The names should be easy to remember and say. What do you think about naming them after outlaws, superheroes, or their rivals?

Chapter 8

Trailblazers—Inspiring Journeys Across Fields of Impact

Vibrant stories fill the pages of history, painting colorful images of those whose lives and adventures continue to light the way for future explorers. Amid these tales lies the heartbeat of genius and inspiring leaders—leaders whose lessons of tenacity and creativity resonate across generations.

Let's explore the lives of some of the most influential people of their time, whose legacy reaches us at present.

Pioneer of Science—Marie Curie

Imagine a time when very few people knew about the mysterious force called radioactivity.

- It was Marie Curie who uncovered its secrets, forever changing the way we understand the world around us.
- Radioactivity opened countless doors for discoveries not just in physics, but also in medicine.
- One of the most amazing gifts from this discovery is the use of radiation to treat severe diseases.
- Marie Curie worked hard alongside her husband, Pierre Curie, in a small laboratory filled with dangerous radioactive substances.

Science During World War I

During World War I, Marie Curie tmade groundbreaking contributions to medical care on the battlefield, particularly in radiography.

- Marie Curie recognized the need for improved medical care for wounded soldiers. She developed mobile X-ray units, known as "Little Curies," that could be transported to the front lines.
- These units allowed for quick imaging and diagnosis of injuries, particularly fractures, helping doctors provide better treatment.
- Marie trained other medical personnel, including doctors and nurses, on how to operate the X-ray machines. She ensured that they had the necessary skills to use the technology effectively in the field.

Alongside her work with X-ray technology, Marie helped raise funds to support her initiatives and the war effort, working with organizations to provide the necessary resources for medical advancements.

- Marie Curie established research facilities dedicated to studying radioactivity.

- These institutions became places where scientists could conduct important experiments and make further advancements in understanding radioactive elements.

- Marie Curie faced many challenges during her lifetime, including working with hazardous materials without safety precautions. It is believed she fell ill due to high exposure to radioactivity.

Two Nobel Prizes!

- Marie Curie was the first woman to win a Nobel Prize—a remarkable achievement in a field traditionally dominated by men.

- She was also the first person to win the Nobel in two different fields: physics (1903) and chemistry(1911).

- Her first award, shared with Pierre Curie and Henri Becquerel, honored their joint research on radioactivity.

- The second prize, awarded to Marie alone, celebrated her continued contributions to chemistry, particularly her work in refining the understanding of radium.

Just think, because of Marie Curie, countless young women decided to pick up test tubes and lab coats instead of feeling they had to stay away from such careers.

What If...?

What if Marie Curie had access to advanced AI technology while conducting her groundbreaking research? How would she use AI to help analyze data and discover new elements, potentially speeding up her experiments? What kind of challenges would she tackle with AI assistance in the fight against diseases?

Incredible Inventions by Nikola Tesla

Nikola Tesla's name might sound like the electric car, but his story is way cooler! Born in 1856, Tesla was an amazing inventor whose ideas changed the world of electricity and technology. Let's take a look at some of his awesome inventions!

Alternating Current (AC)

Imagine a world where every home had to be right next door to a power plant to keep the lights on. Sounds strange, right?

- Before Tesla came along, people used something called direct current (DC) for electricity, which couldn't travel very far.
- But Tesla changed everything! He created an alternating current (AC), an incredible discovery that allowed electricity to travel long distances.
- Thanks to AC, we enjoy electric lights and gadgets in our homes even if the power plants are miles away.
- Tesla's AC system became the standard all over the world, shaping the modern world as we know it.

But wait, there's more! Tesla didn't stop at making electricity better; he also dreamed of a world without wires.

Wireless Communication

In the late 1800s, when most people couldn't even imagine Wi-Fi or wireless communication, Tesla was already experimenting with sending electricity through the air!

- His early work helped pave the way for today's wireless technologies, like cell phones and radio waves.

- When Tesla lived, communication means were wired: the telegraph and telephone lines. Tesla saw this future long before it happened.

- He experimented with high-frequency currents and wireless transmission technology.

- Tesla's work laid the foundation for the use of radio waves in communication. He believed in the potential of transmitting signals through the air, which would revolutionize communication methods.

- In 1893, Tesla conducted public demonstrations of wireless transmission of energy and information. He couldn't imagine how far his ideas could reach!

Tesla Coil

Have you ever seen those cool science demonstrations with big sparks of electricity jumping through the air? That's all thanks to the Tesla coil!

- Tesla invented this device to produce high-voltage electricity.

- The Tesla coil became famous because it made electricity visible.

- Even today, it is used in radios and TVs, showing just how ahead of his time Tesla was.

Tesla's Adventures and Misfortunes

- After studying engineering in Europe, he moved to America with only a few cents in his pocket.

- He worked with Thomas Edison, another famous inventor, but they had different ideas about electricity.

- While Edison focused on making money, Tesla loved pushing boundaries with his inventions.

- Eventually, they parted ways, and Tesla went on to create even more amazing things, like designs for X-rays and remote controls.

Despite being a genius, Tesla faced countless obstacles during his life, including financial struggles. This prevented him from completing many of his experiments. Still, he envisioned the world we live in today!

What If...?

What if Nikola Tesla could travel to the present and see the many applications of his ideas about wireless communication and energy transmission? These are the basis for virtual reality and other immersive experiences that allow us to explore other virtual realities. What innovative projects would he create if he had access to AR and VRP?

Adventures of Amelia Earhart

Amelia Earhart wanted to see the world... from above! She dreamed of traveling around the world on her yellow airplane.

First Solo Traveler

- In 1932, Amelia Earhart made history as the first woman to fly solo across the Atlantic Ocean—a remarkable achievement for the time.

- She piloted her red Lockheed Vega 5B from Harbor Grace, Newfoundland, to Londonderry, Northern Ireland.

- This journey took about 15 hours and demonstrated not only her skills but also a profound statement on women's capabilities in aviation.
- Before Earhart's accomplishment, such endeavors were almost exclusively male-dominated, leaving little space for women to explore their potential.

Other Amelia's Records

- Amelia earned a reputation as a record-breaking aviator, pursuing numerous challenging flights with tenacity.
- In August 1932, shortly after her transatlantic success, she made the first solo nonstop flight by a woman across the United States, traveling from Los Angeles to Newark, New Jersey.
- She established a women's distance record of over 2,400 miles and also broke the previous women's speed record.

A Mysterious Disappearance

- Accompanied by navigator Fred Noonan, she aimed to cover approximately 29,000 miles, a journey that promised adventure and discovery.
- However, the pair never completed their flight, and despite extensive search efforts, their fate remains one of the great unsolved mysteries of the 20th century.
- In 1937, while attempting a monumental round-the-world flight, Amelia vanished over the Pacific Ocean.
- The intrigue of her disappearance is deeply tied to her legacy of adventure. Many see it as emblematic of the inherent risks associated with exploration and pushing boundaries.

Other Achievements

- Amelia Earhart was not just a pioneer in the sky; she was also a passionate advocate for women in aviation.
- She co-founded the Ninety-Nines, an organization dedicated to providing support and opportunities for female pilots.
- As the first president of this group, Earhart helped empower future generations, encouraging women to take control of their destinies within the soaring industry of aviation.

What If...?

What if Amelia Earhart's final adventure took her to uncharted islands, where she encountered unique wildlife and discovered hidden treasures? How would she navigate through storms and unexpected challenges, using her skills as a pilot and explorer?

Roald Amundsen, the Intrepid Explorer

- Roald Amundsen was a famous Norwegian explorer who made significant contributions to the exploration of the polar regions.
- Born on July 16, 1872, Amundsen grew up in a family that revered the sea.
- He was close to the maritime world since he was a child, and this fueled his interest in exploration.
- Amundsen joined a Belgian Antarctic expedition in 1897 and became interested in the extreme conditions in polar regions.

After this expedition, he realized that he wanted to lead his own journey to the icy south.

First Expedition to the North

- In 1903, he set sail for the Northwest Passage, a sea route that connects the Atlantic and Pacific Oceans through the Arctic Ocean.

- This passage was long sought after by explorers but had proven difficult to navigate due to ice and harsh weather.

- Amundsen and his crew faced various challenges during the journey, but in 1906, they successfully completed the passage, making Amundsen one of the few explorers to do so.

Toward the South!

After navigating the Northwest Passage, Amundsen turned his attention to Antarctica.

- In 1910, he embarked on an ambitious journey that aimed to reach the South Pole.

- His ship, the Fram, was specially built for polar exploration, allowing it to withstand the ice.

- The journey to the South Pole started in 1911, and Amundsen's team faced extreme cold, fierce winds, and blinding snowstorms.

- On December 14, 1911, Amundsen and his team reached the South Pole, becoming the first people ever to do so.

- Upon arriving at the pole, they planted the Norwegian flag and built a small tent. Instead of claiming the land for Norway, Amundsen left a note acknowledging their achievement.

The North Pole

In the years that followed his triumphant journey to the South Pole, he sought new adventures. He turned his attention to the Arctic and aimed to reach the North Pole.

- He embarked on a new expedition in 1926, where he successfully crossed the Arctic by airship.
- Amundsen's exploration spirit was also marked by a sense of competition. He was aware that others aimed to achieve similar goals.
- Notably, American explorer Robert Peary claimed to have reached the North Pole in 1909.
- Amundsen, however, remained focused on his mission rather than engaging in a race to claim credit.

His Last Feat

- Roald Amundsen's life ended in 1928 while on another expedition.
- He was attempting to rescue fellow explorer Umberto Nobile, whose airship had crashed in the Arctic.
- Amundsen disappeared during the rescue mission, and despite extensive search efforts, his body was never found.

This tragic end did not overshadow his accomplishments. Instead, it solidified his status as a significant figure in the history of exploration.

What If...?

What if Roald Amundsen, during his remarkable adventures to the South Pole, discovered a hidden land beyond the ice of Antarctica, filled with extraordinary creatures and lush landscapes? How would he and his team react to this exciting discovery, and what kind of new adventures would they embark on in this uncharted territory?

Valentina Tereshkova, a Woman Out of This World

- Valentina Tereshkova was born on March 6, 1937, in a small village in Russia.
- As a young girl, she would often look up at the sky and dream of traveling among the stars.
- After finishing school, Valentina began her training as a parachutist. She quickly excelled in this sport, making over 150 jumps.
- Her skills and dedication caught the attention of the Soviet space program, which was looking for candidates for their next big mission.

The Space Race

- The program was searching for someone who could endure the physical and psychological challenges of space travel.
- Valentina's experience as a parachutist made her an ideal candidate.
- In 1962, Valentina was selected from over four hundred applicants to be the first woman in space.
- This historic selection showcased the Soviet Union's commitment to equality and progress.

The Mission to Space

- On June 16, 1963, Valentina Tereshkova made history as the first woman to fly in space aboard the Vostok 6.
- This mission lasted nearly three days, during which she orbited the Earth forty-eight times.
- During the flight, she conducted experiments and took photographs of the Earth's surface contributing valuable data for future space missions.
- Valentina's mission was significant not just because she was a woman; it also demonstrated the advancements in technology and the potential for space exploration.

Her success broke significant barriers and inspired people worldwide. After the mission, Valentina became an international symbol of women's achievements in science and space.

What If...?

What if Valentina Tereshkova, the first woman to fly in space, embarked on a new mission today as a leader in space exploration? What if her mission involved training a diverse team of young astronauts to explore Mars and promote international cooperation in space travel?

Chapter 9

Nature's Secrets

Discovering the wonders of nature is like opening a door to a world full of fascinating secrets. From oceans teeming with life to skies filled with fluttering butterflies, each corner of our planet hides remarkable stories waiting to be told.

Join us as we unravel these intriguing tales from the wild, sparking curiosity and a deeper appreciation for the wonders of the natural world.

Camouflage Abilities of Octopuses

In the vast, mysterious ocean depths, octopuses have mastered the art of camouflage, showcasing remarkable adaptations that turn them into escape artists of the sea.

- These fascinating creatures can transform their appearance almost instantly. This skill is crucial for survival in an environment filled with threats.
- An octopus's skin is equipped with specialized cells known as chromatophores, which they use to change colors swiftly.
- Imagine tiny balloons filled with colored pigments spread across their skin; these expand or contract to create vibrant patterns matching their surroundings.
- This color-changing ability allows octopuses to blend perfectly with corals, rocks, or sandy seabeds, making them nearly invisible to lurking predators and unsuspecting prey.

The Master of Disguise

But octopuses don't stop at changing colors.

- They possess another trick up their sleeves—texture transformation.
- With the help of muscle-controlled projections called papillae on their skin, octopuses can alter their skin texture to match their surroundings.
- This ability allows them to replicate the rough surface of a rock, the smoothness of coral, or even the fine grains of sand.
- Their camouflage capabilities allow them to hide in plain sight within diverse underwater landscapes.

Red Light, Green Light!

- Octopuses can also enhance their disguise through behavior.

- Often, they remain motionless, mimicking the stillness of their surroundings to avoid detection.
- In some cases, they adopt mimicry as an advanced strategy.
- The mimic octopus takes this skill to a whole new level—it impersonates other marine species.
- This intelligent octopus imitates not only the appearance but also the movement patterns of more dangerous or weaker animals, confusing predators or threatening rivals.

This helps octopuses put food on their plates every day! But... why do octopuses go to such great lengths to disguise themselves?

- The answer lies in the harsh realities of ocean life, where being stealthy could mean evading a predator's keen eye or successfully ambushing unsuspecting prey.
- Octoposus have developed these adaptations over millions of years.

What If...?

What if other animals had the same ability as octopuses to camouflage? How would that change life in the depths of the ocean? If you had the octopuses' ability to disguise, what strategy would you use: color, shape, or motion?

Monarch Butterflies, Seasonal Travelers

Have you ever heard of a butterfly that takes an incredible journey across entire continents?

- Monarch butterflies are famous for their long-distance migration, traveling thousands of miles from North America all the way to Mexico.

- Now, what's really fascinating is that this amazing trip isn't completed by just one butterfly but by several generations!
- Imagine starting a journey and passing the baton to your kids and then their kids' kids before reaching the finish line. That's what monarchs do.

How Do They Plan the Trip?

Speaking of marvels, did you know that sensory cues play a huge role in monarch migration too?

- These spectacular butterflies use something almost magical to find their way—solar positioning.
- It is like having an internal compass that guides them through the sky, relying on the sun's position.
- They respond to environmental signals that help them decide when to start their journey or when it's time to turn back.
- It is like reading cosmic road signs! These cues are part of what powers their astonishing biannual voyage across the continent.

A Challenging Adventure

But let's pause and think about what they face during this epic travel. The journey isn't a walk in the park.

- Monarchs have to deal with severe weather conditions, dodge predators, and navigate through landscapes that change over time.
- Monarch populations are declining due to habitat loss, which is quite worrying. Forested areas continue to disappear year after year.

- Picture waking up one day and finding half of your neighborhood vanished, with no clear path home. That's happening to monarchs, with crucial migratory routes being lost.

Monarch butterflies and other migrant species of birds flying south in autumn as well as whales breaching in the ocean, tell us about the rhythm of nature.

What If...?

What if monarch butterflies set out on their incredible migration south but faced unexpected challenges along the way, like strong winds and changing weather patterns that made it hard to find their routes? How would they rely on their instincts and teamwork to navigate through unfamiliar landscapes, sharing stories and tips to stay on course?

Bioluminescent Creatures

Imagine walking on a warm summer night and seeing the twinkling of fireflies lighting up your path. The light comes from some living creatures!

- Their glow comes from an amazing natural process called bioluminescence, where living organisms produce light through chemical reactions in their bodies.

- This fascinating phenomenon plays a role in predation, defense, and communication among various species.

The Chemistry Behind the Magic

- Bioluminescence is driven by two chemicals: luciferin and luciferase. When these two interact, they produce light.

- Think of luciferin as the fuel and luciferase as the engine that ignites it.

- This reaction can happen in diverse settings with captivating results.

Bright Species

- Marine life, for example, showcases many glowing wonders.
- The ocean, often seen as vast and mysterious, houses jellyfish and deep-sea fish that dazzle with pulsating lights.
- These illuminations can confuse predators, attract prey, or even communicate warning signals to other marine creatures.
- On land, fireflies create stunning displays in mid-air, while certain fungi offer an eerie glow on the forest floor.

Scientific Applications of Bioluminescence

Further applications in biotechnology hold promise!

- Imagine crops that could be engineered to glow when they need water, helping farmers manage their fields more efficiently.
- Or bioluminescent trees illuminating city streets, reducing electricity demand.
- Certain algae and bacteria serve as indicators of environmental changes in the sea, pollution in the air, and changes in water quality.

What If...?

What if bioluminescent animals could be used in amazing ways to help people and the planet? How would scientists and engineers work together to harness the light from glowing jellyfish or fireflies to create natural streetlights that save energy and brighten up dark areas? What if bioluminescent creatures could help us detect pollution in oceans by changing their glow when exposed to harmful chemicals?

Axolotls

What is the weirdest animal you have ever seen? These are among the most creatures in the animal kingdom. Allow us to introduce you to the axolots!

- Axolotls are fascinating creatures that belong to a group of amphibians known as salamanders.
- They are unique because they spend their entire lives in water.
- They typically have a soft body with a wide head and fringed external gills that resemble feathery appendages.
- Their colors can vary, ranging from dark brown to pink or white, creating an appealing aesthetic that many find endearing.

Forever Young

- Unlike many other amphibians, axolotls keep their juvenile features throughout their lives.
- This process is called neoteny, which means they retain their gills and tail structure.
- Instead of becoming adults that venture onto land, axolotls remain aquatic, which is one reason they attract so much attention.

Where Do They Live?

- These creatures are often found in the lake complex of Xochimilco in Mexico.

- This habitat consists of canals, rivers, and lakes, which offer a rich environment for axolotls.

Axolots are wonderful pets!

- For those interested in keeping an axolotl as a pet, it's essential to know how to create a suitable environment.

- A proper aquarium is crucial, as axolotls require plenty of space to swim.

- A tank of at least 20 gallons is recommended for a single axolotl. Make sure to include a filtration system to keep the water clean, as axolotls are sensitive to water quality.

- The temperature of the water should ideally be between 60 to 68 degrees Fahrenheit to mimic their natural habitat.

- Fine sand or smooth rocks can be a good choice to prevent axolotls from accidentally ingesting pebbles.

- A well-decorated aquarium can help simulate their natural environment and make them feel more secure.

What If...?

What if they discovered that, unlike cats and dogs, axolotls don't need walks or cuddles but thrive in a carefully maintained aquarium with clean water and special food? How might you enjoy watching their axolotls swim and change colors, while also appreciating the playful antics of their other pets?

Spectacular Atmospheric Phenomena

Atmospheric phenomena are fascinating events that occur in the sky and can influence weather, climate, and even our daily lives.

Atmospheric Optics: Sun Dogs and Halos

Atmospheric optics is another area of atmospheric phenomena that presents a range of beautiful visual effects.

One of these effects is the sun dog:

- A sun dog is a bright spot that appears on either side of the sun.
- Sun dogs occur when sunlight refracts through ice crystals in the atmosphere, creating bright flickering spots.
- You are more likely to see sun dogs during cold weather when ice crystals are present in the upper atmosphere.

Halos are another form of atmospheric optics.

- They typically appear as rings around the sun or moon.
- Like sun dogs, halos also occur due to the refraction of light through ice crystals.
- To spot a halo, look for a circular glow around the sun on cold, clear days.
- This natural light display can be awe-inspiring and provides an opportunity to share your observations with others.

Fascinating Ice Phenomena

It is interesting to look at the various ways that ice can form and the unique phenomena that can arise from it.

Ice Crystals

- When water vapor in the air cools and freezes, it forms tiny crystals.

- These ice crystals can come together to make snowflakes. Each snowflake has a unique structure, which is why they look so different from one another.

- Ice crystals can also form on surfaces when the temperature is low, creating frost.

- This happens when water vapor in the air comes into contact with cold surfaces and freezes.

- To see frost, you can simply check your car's windshield in the early morning during winter.

Icebergs

- Icebergs are massive pieces of freshwater ice that have broken off from glaciers or ice shelves.

- These massive blocks can be found floating in the ocean, and their size is astonishing.

- They are typically white or blue, depending on how dense the ice is and how much air it contains.

- The icy blue color comes from light being absorbed, and the less dense ice scattering the blue light.

- Sometimes, icebergs can flip over, revealing their ice underside, which can be a fascinating sight.

Glaciers

- They are large masses of ice that form from compacted snow over many years.

- Glaciers move slowly down mountains and valleys, shaping the landscape as they go.

- This movement can create valleys, lakes, and even fjords.

- Many national parks offer guided tours where you can see glaciers up close.

What If...?

What if you woke up one morning to discover a sky filled with colorful clouds that danced and changed shapes, creating stunning patterns? How would you feel as you stepped outside and witnessed shimmering rainbows or glowing orbs floating above your head?

Chapter 10

Techno Whiz-Bangs

With each passing day, artificial intelligence (AI) and advanced technologies become more integrated into our daily lives, offering solutions that were once thought to be the realm of science fiction. This incredible technological advance is transforming industries ranging from manufacturing and healthcare to education and beyond.

Get ready to embark on an eye-opening journey through the world of "Techno Whiz-Bangs," where the only limit to what we can achieve is the breadth of our imagination.

The Rise of Artificial Intelligence

Imagine being able to solve problems just like a human but with the power of machines. That's what AI does, and it feels almost magical!

- Every day, AI is changing how we interact with digital platforms by providing personalized experiences.
- Have you ever noticed recommendations popping up on your favorite streaming service or app? That's AI at work!
- It is learning about our likes and dislikes to tailor our experiences in a way that suits us best.

AI in the Classroom

When it comes to education, AI is playing a game-changing role.

- Think about a classroom where each student gets lessons crafted just for them.
- With AI-powered programs, students can learn at their own pace, focusing on areas they find tricky while zipping through what they already know.
- It is like having a personal tutor available 24/7!

At the E.R.

AI is assisting doctors and nurses in amazing ways.

- From reading medical scans with precision to predicting disease outbreaks, AI is helping healthcare professionals make smarter decisions.
- For example, some AI tools can now analyze radiology images to detect diseases more accurately than ever before.
- This technology aids in planning effective treatments, thereby saving lives.

- Even though there are challenges, such as ensuring data privacy and understanding complex AI models, the potential benefits are enormous.

Robots and AI

- As AI continues to evolve, it may lead to robots that can perform tasks that require decision-making and adaptability, much like humans.
- While traditional robots operate based on predefined instructions, integrating AI enables robots to adapt and make decisions in real-time.
- Imagine robots helping in various fields, like agriculture, industrial cleaning, and even disaster response.
- AI techniques, such as machine learning and deep learning, enable robots to learn from their experiences and improve their functionalities.
- For example, a robot can recognize objects in its environment and adapt its behavior accordingly.

The possibilities are endless, and who knows what groundbreaking developments might happen next?

What If...?

What if robots could take on chores at home, like cleaning, cooking, or even gardening, allowing families to spend more quality time together? What if we had friendly robots as companions that could help us practice sports or learn new skills, encouraging us to be our best selves? Would you like to have a friend robot?

How 3D Printers Are Changing Industries

Have you ever wished to create anything you wanted layer by layer, almost like building a sandwich, but way cooler? That's the magic of 3D printing technology.

- This incredible tech lets us design and make objects right in front of our eyes, revolutionizing how things are made across many industries.

How Do 3D Machines Work?

- Imagine a 3D printer as a fancy machine that builds things from the bottom up, one thin layer at a time.
- It is like crafting, but instead of using tools or hands, the printer uses special materials to form each piece based on a digital plan.
- With advanced art supplies, you can shape almost anything, from toys to tools, just by following instructions!

A Revolution in Industry

- Traditionally, making something new meant a lot of time and expenses, especially when only a few specific items were needed.
- With 3D printing, companies can produce items quickly and cost-effectively.
- For example, imagine needing a special part for a car or airplane—rather than waiting weeks for it to be shipped from across the globe, manufacturers can now print it right where it's needed.
- This saves not only time but also money, and it allows for the creation of personalized products. It's like ordering a custom pizza delivered hot and fresh just for you!

Technology to Help Healing

Moving into healthcare, the possibilities with 3D printing are both exciting and heartwarming.

Surgeons and doctors use this technology to craft custom prosthetics and surgical models that fit patients just right.

Previously, getting a prosthetic limb involved a long wait and might not always provide the perfect fit. Now, with 3D printing, a person can have a tailor-made prosthetic that fits them perfectly.

Hospitals even use detailed 3D prints of hearts, bones, and other body parts to plan surgeries.

Other Applications

- In educational settings, 3D printing has really taken off. For instance, universities sometimes use this tech to produce precise replicas of rare historical artifacts or complex scientific models.

- It means students can handle and study them closely, enriching their understanding of history or science in ways textbooks can't match.

- In some classes, students even get to design and print their own projects, bringing imagination to a tangible form—a thrilling combination of learning and creating.

But what about the future of this technology? As 3D printing continues to evolve, its impact will likely grow.

- New kinds of materials are being developed that will allow these printers to create things we haven't even dreamed of yet.

- Imagine clothes printed from smart materials that change designs with your mood, or printing entire buildings layer by layer in just days!
- Fashion designers are experimenting with intricate outfits that couldn't be sewn by hand, pushing the boundaries of what's possible in style and attire.

The best part? This technology invites endless exploration and experimentation, empowering creators everywhere to think outside the box—or the printed mold!

What If...?

What if 3D technology became a magical tool that allowed you to create anything? How would you use 3D printing to bring your wildest dreams to life? What if 3D simulations let you explore ancient civilizations and distant planets without ever leaving your classroom, learning through real-life experiences?

The Evolution of Computers and the Internet

Once upon a time, people lived without the Internet, but that is part of a past that won't go back. Computers and the internet have reshaped the way we live, work, and communicate. But how did we get here?

The First Computers

- Our story begins with early computers, which looked nothing like the sleek devices we're familiar with today.
- In the early days, computers were large mechanical devices used primarily for calculations in academic settings.

- The first electronic computers emerged in the mid-20th century, such as the Electronic Numerical Integrator and Computer (ENIAC), which took up entire rooms!
- These machines laid the groundwork for the electronic systems that power modern computing.
- Slowly, over the years, massive machines evolved into the compact and efficient devices we carry in our pockets.

Shrinking and Improving

- One of the most significant milestones was the development of microprocessors in the 1970s.
- A microprocessor is a tiny chip that acts as the brain of a computer, performing tasks and running programs.
- This leap in technology led to the era of personal computing.
- Computers became smaller, more affordable, and accessible to ordinary people.
- Companies like Apple and IBM played vital roles in bringing personal computers like the Apple II and IBM PC into homes and offices around the globe.
- These personal computers transformed the way individuals interacted with technology, marking the beginning of an age where computers started to become household items.

With Our Computer to Anywhere

But the revolution didn't stop there. Enter mobile technology.

- With the advent of laptops, smartphones, and tablets, computing became truly portable. We could now carry our digital worlds with us wherever we went!
- This mobility changed how we communicate, access information, and stay connected with one another, blurring the lines between work, home, and social lives.

The Global Net www.

Parallel to the rise of computing was the development of the internet—a phenomenon that has had the largest impact on global communication since the invention of the telephone.

- Starting as a project called ARPANET in the 1960s, the aim was simple: connect a series of computers across different locations.
- By the late 1980s and early 1990s, this network evolved into what we now know as the World Wide Web: www.
- Browsers like Netscape Navigator and Internet Explorer made web access easier, and the Internet expanded into homes and businesses.
- It wasn't long before email became a staple mode of communication, e-commerce boomed, and online education began providing learning opportunities beyond geographical boundaries.
- The advent of wireless communications (thank Tesla!) freed users from cables and allowed them to connect to the internet from nearly anywhere.
- Soon, we had innovations like Wi-Fi and 4G LTE.

People from every corner of the earth can now share ideas, cultures, and inventions at the click of a button.

What If...?

What if computers and the internet evolved into powerful tools that could connect our minds and imaginations directly to the digital world? How might this evolution change the way people share ideas and solve problems, sparking new inventions that improve our lives?

Conclusion

We are close to the end of this incredible journey through the fascinating realms of science, history, art, and so much more. Let's take a moment to reflect on the amazing things we've discovered.

You've learned about extraordinary creatures like octopuses, which can change colors in an instant to blend into their surroundings. Isn't it mind-boggling how nature has crafted such unique abilities? Or perhaps you were intrigued by the evolution of soccer rules, telling us how this beloved game has transformed over time into what we enjoy today. These stories are more than just fun facts; they show us how curiosity drives discovery and how each new piece of knowledge enriches our lives.

Think back to some of your favorite parts from the book. Maybe the dazzling dance of the Northern Lights captured your imagination, or maybe learning about ancient civilizations and their mysteries kept you flipping pages late into the night. Every chapter was a doorway into a different world, revealing the secrets of our universe and sparking that little flame of curiosity inside you. This is the magic of exploration and learning—how it makes every day an adventure filled with endless possibilities.

But this book is only the beginning of your quest for knowledge. If you've been amazed by what you've read, there's so much more waiting out there for you to discover!

Let's not forget the power of curiosity itself—a key theme throughout this book. Curiosity has fueled the achievements of many inspiring individuals, from inventors to scientists. Do you remember Marie Curie? Her relentless curiosity pushed her to explore radioactivity, leading to groundbreaking discoveries that changed science forever. Who knows where your own questions might lead you one day? Perhaps you'll be the one to solve a big problem or make a discovery that others could never have imagined.

While reading this book, you may have felt a spark of wonder about the vastness and diversity of our world. Hold onto that feeling, keep feeding it, and let it guide you throughout your life. There are countless wonders in our everyday environment, waiting to be noticed and appreciated. From the intricate patterns of leaves to the songs of different birds, observing these details encourages a deep sense of connection with nature and culture.

Even as you move forward, always remember that the most incredible discoveries of all might come from within yourself. Embrace the challenge of learning something new every day, asking questions, and cultivating a lifelong sense of awe at the marvels that surround us.

By nurturing your curiosity, you not only enrich your life but might also inspire others to embark on their own journeys of discovery. Share what you've learned with friends and family—become someone who sparks conversations and ignites interests, just like the fun, conversation-starting facts you've collected along the way.

Armed with the knowledge you've gained and the curiosity you've honed, there's no limit to what you can achieve. So go forth, young explorers, and leave no stone unturned in your quest to uncover the secrets of our world. The journey has only just begun, and who knows what incredible things await you beyond these pages? Go out there, let your curiosity lead, and be ready for amazing discoveries!

References

African Masks: The rich cultural heritage and artistic significance. (n.d.). Berj Art Gallery. https://www.berjartgallery.com/news/african-masks-the-rich-cultural-heritage-and-artistic/

Algeria. (n.d.). The British Museum. https://africanrockart.britishmuseum.org/country/algeria/

American Hospital Association. (2022, June 7). 3 Ways 3D printing is revolutionizing health care. American Hospital Association. https://www.aha.org/aha-center-health-innovation-market-scan/2022-06-07-3-ways-3d-printing-revolutionizing-health-care

Art and Culture. (n.d.). Kanaga Africa Tours. https://www.kanaga-at.com/en/trip-info/algeria-en/art-and-culture/

Axolotl. (n.d.). National Geographic. https://www.nationalgeographic.com/animals/amphibians/facts/axolotl

Bahcall, N. (2015, March 17). *Hubble's Law and the expanding universe.* Proceedings of the National Academy of Sciences. https://doi.org/10.1073/pnas.1424299112

Belmessous, S. (2013, March 13). Assimilation against colonialism. In *Assimilation and Empire.* Oxford Academic. https://doi.org/10.1093/acprof:oso/9780199579167.003.0004

Bioluminescence. (2023, October 19). National Geographic. Education.nationalgeographic.org. https://education.nationalgeographic.org/resource/bioluminescence/

Bold Himalaya. (2021, January 5). Bold Himalaya. https://boldhimalaya.com/blog/mount-everest-facts-and-information-90

Bomzer, R. (2022, September 19). 50+ unusual musical instruments. Carved Culture. https://www.carvedculture.com/blogs/articles/unusual-musical-instruments?srsltid=AfmBOorGU2YwEm3yk_TFA6s-zvqoKqtQVpYy1Majt52oOT4dBZkULv8k

Breen, K. (2024, July 1). Here's the full list of hurricane names for the 2024 season. CBS News. https://www.cbsnews.com/news/hurricane-names-list-2024-season/

Brinkhof, T. (2023, December 19). *"Phantom time hypothesis": Did a power-hungry pope fabricate centuries of history?.* Big Think. https://bigthink.com/the-past/phantom-time-hypothesis/

Captain Methew. (2023, September 20). *The Bermuda Triangle mystery: A modern maritime Perspective.* The Maritime Post.

https://themaritimepost.com/2023/09/bermuda-triangle-mystery-conspiracy/

Chen, L. (2023, August 2). Influence of music on the hearing and mental health of adolescents and countermeasures. *Frontiers in Neuroscience, 17*. https://doi.org/10.3389/fnins.2023.1236638

Cleaver, G. (2019, November 22). Multiverse theories: Philosophical and religious perspectives. In *Oxford Research Encyclopedia of Religion.* https://doi.org/10.1093/acrefore/9780199340378.013.157

Cochrane, D. (2022, May 20). *90 Years After Her Solo Transatlantic Flight: What Would Amelia Earhart Think About Women in Aviation?* National Air and Space Museum. https://airandspace.si.edu/stories/editorial/90-years-Earhart-solo-transatlantic-flight

Cooper, H. (2021, November 15). *Illuminating the facts of deep-sea bioluminescence.* Monterey Bay Aquarium. https://www.montereybayaquarium.org/stories/bioluminescence

*Dark energy and dark matte*r. (n.d.). Center of Astrophysics. https://www.cfa.harvard.edu/research/topic/dark-energy-and-dark-matter

Dave, M., & Patel, N. (2023, May 26). Artificial intelligence in healthcare and education. *British Dental Journal, 234*, 761–764. https://doi.org/10.1038/s41415-023-5845-2

Davenport, T., & Kalakota, R. (2019, June). The potential for artificial intelligence in healthcare. *Future Healthcare Journal, 6*(2), 94-98. https://doi.org/10.7861/futurehosp.6-2-94

Earthquakes and tsunamis: Facts, FAQs, and how to help. (n.d.). World Vision. https://www.worldvision.org/disaster-relief-news-stories/earthquake-tsunami-facts

The Editors of Encyclopedia Britannica. (n.d.). Bermuda Triangle. In *Encyclopedia Britannica.* Retrieved on October 26, 2024. https://www.britannica.com/place/Bermuda-Triangle

The Editors of Encyclopedia. (n.d.). Marie Curie. In *Encyclopedia Britannica.* Retrieved on October 26, 2024. https://www.britannica.com/biography/Marie-Curie.

The Editors of Encyclopedia Britannica. (n.d.). Roald Amundsen. In *Encyclopedia Britannica.* Retrieved on October 26, 2024. https://www.britannica.com/biography/Roald-Amundsen.

The Editors of Encyclopedia Britannica. (n.d.). Sun dog. In *Encyclopedia Britannica.* Retrieved on October 26, 2024. https://www.britannica.com/science/sun-dog

The Editors of Encyclopedia Britannica. (n.d.). Valentina Tereshkova. In *Encyclopedia Britannica.* Retrieved on October 26, 2024.https://www.britannica.com/biography/Valentina-Tereshkova.

Fienberg, R. (2006, July 25). *What is a Sundog, and how did "sundogs" get their name?.* Sky & Telescope. https://skyandtelescope.org/astronomy-resources/astronomy-questions-answers/why-are-sundogs-called-by-that-name/

First woman in space: Valentina. (2013, June 16). The European Space Agency. https://www.esa.int/About_Us/ESA_history/50_years_of_humans_in_space/First_woman_in_space_Valentina

Fung, B. (2024, May 10). *Why tonight's massive solar storm could disrupt communications and GPS systems.* CNN. https://www.cnn.com/2024/05/10/business/sunspots-disrupt-phones-gps-scn/index.html

Gantley, M. (2023, February 27). *Stonehenge, the enigmas of the stone circle.* National Geographic. https://historia.nationalgeographic.com.es/a/stonehenge-enigmas-circulo-piedra_8857

Guerra, P. A. (2020, December). The monarch butterfly as a model for understanding the role of environmental sensory cues in long-distance migratory phenomena. *Frontiers in Behavioral Neuroscience, 14.* https://doi.org/10.3389/fnbeh.2020.600737

Gurung, P. (2024, January 31). *Mount Everest: The highest peak in the world.* Discovery World Trekking. https://www.discoveryworldtrekking.com/blog/mount-everest

History.com Editors. (2018, August 21). *Atlantis.* HISTORY; A&E Television Networks. https://www.history.com/topics/folklore/atlantis

The history of soccer. (2024). Prezi. https://prezi.com/p/wrwl0uv0yx_g/the-history-of-soccer/

Hossain, A. Admin, (2023, October 8). *Wormholes: The cosmic tunnels.* The Journal of Young Physicists. https://www.journalofyoungphysicists.org/post/wormholes-the-cosmic-tunnels

Hossenfelder, S. (2024, October 14). Why the multiverse is religion, not science. *Back Reaction Blogspot.* http://backreaction.blogspot.com/2019/07/why-multiverse-is-religion-not-science.html

Hughes, A., Kirksey, E., Palmer, B., Tivasauradej, A., Changwong, A.A. & Chornelia, A. (2023, November). Reconstructing cave past to manage and conserve cave present and future. *Ecological Indicators, 155.* https://doi.org/10.1016/j.ecolind.2023.111051

Hunt, I. W. (n.d.). Nikola Tesla. In *Encyclopedia Britannica.* Retrieved on October 24, 2024. https://www.britannica.com/biography/Nikola-Tesla

The iconic Olympic Torch: History, symbolism, and tradition. (n.d.). The Olympic Torch https://www.runnersneed.com/the-olympic-torch.html

Kennedy, L. (2021, November 19). *The Native American origins of Lacrosse.* HISTORY. https://www.history.com/news/lacrosse-origins-native-americans

Kennell, J. (2015, November 18). *The 8 wackiest weather phenomena on Earth.* The Science Explorer. https://www.thescienceexplorer.com/the-8-wackiest-weather-phenomena-on-earth-406

Kirigha, F. (2023, September 12). *Introduction: Unveiling the artistry and heritage of African mask craft.* Tulia African Store. https://www.tuliaafricanstore.com/post/introduction-unveiling-the-artistry-and-heritage-of-african-mask-craft?srsltid=AfmBOooEZzY4XcvTvpH7lAX3lWxrrSyt4DlokEgwBtM3klpl73r0Z9W1

Leiner, B., Cerf, V., Clark, D.D., Kahn, R.E., Kleinrock, L., Lynch, D.C., Postel, J., Roberts, L.G., & Stephen WolffInter, S. (1997). *Brief History of the Internet.* Internet Society. https://www.internetsociety.org/internet/history-internet/brief-history-internet/

MacEachern, S. (2000, June). Genes, tribes, and African history. *Current Anthropology, 41*(3). https://doi.org/10.1086/300144

Marcinkowski, C. (2010, July). Herbert Illig-Who has tampered with the clock? How 300 years of the Middle Ages were invented. *Islam and Civilisational Renewal, 1*(4). https://go.gale.com/ps/i.do?id=GALE%7CA270895992&sid=googleScholar&v=2.1&it=r&linkaccess=abs&issn=2041871X&p=AONE&sw=w&userGroupName=anon%7E77f53f92&aty=open-web-entry

Marie Curie. (2024, March 5). Biography. https://www.biography.com/scientists/marie-curie

Measuring the universe's expansion rate. (n.d.). HubbleSite. https://hubblesite.org/mission-and-telescope/hubble-30th-anniversary/hubbles-exciting-universe/measuring-the-universes-expansion-rate

Medal, Amelia Earhart, first woman to cross the Atlantic by airplane. (n.d.). National Air and Space Museum. https://airandspace.si.edu/collection-objects/medal-amelia-earhart-first-woman-cross-atlantic-airplane/nasm_A19640152000

Meyer, F (n.d.). *How octopuses and squids change color.* Smithsonian. https://ocean.si.edu/ocean-life/invertebrates/how-octopuses-and-squids-change-color

Mhmosharrf. (2024, March 20). *The evolution of football: From traditional to modern.* Medium. https://medium.com/@mhmosharrf/the-evolution-of-football-from-traditional-to-modern-79ca2ec9ec4c

Muckerman, A. (2023, June 5). *The fascinating new theory on the function of Stonehenge, England's most mysterious and famous monument.* BBC. https://www.bbc.com/mundo/vert-tra-65806110

Native American Sports. (n.d.). In *Encyclopedia.com.* https://www.encyclopedia.com/history/news-wires-white-papers-and-books/native-american-sports

Nepilova, A. (2024, August 19). *'Somewhere between mesmerising and terrifying': ten of the weirdest musical instruments out there.* Classical Music. https://www.classical-music.com/features/instruments/weirdest-instruments

Nikola Tesla - Inventions, quotes and death. (2024, February 6). Biography. https://www.biography.com/inventors/nikola-tesla

Nijhuis, M. (2023, December 14). *Follow the monarch on its dangerous 3,000-mile journey across the continent.* National Geographic. https://www.nationalgeographic.com/premium/article/monarch-butterfly-migration-endangered

Octopus Adaptations for Survival. (n.d.). Seafood Peddler. https://www.seafoodpeddler.com/octopus-adaptations-for-survival/

Olatunji, G., Osaghae, O. W., & Aderinto, N. (2023, October). Exploring the transformative role of 3D printing in advancing medical education in Africa; A review. *Annals of Medicine and Surgery, 85*(10), 4913-4919. https://doi.org/10.1097/ms9.0000000000001195

The Olympic Symbol and other elements of the Olympic Identity. (n.d.). International Olympic Committee. https://olympics.com/ioc/faq/olympic-symbol-and-identity/what-is-the-olympic-flame-and-torch-relay

The origins of the piano: The story of the piano's invention. (n.d.). Yamaha. https://www.yamaha.com/en/musical_instrument_guide/piano/structure/

Peralta, L. (2022, May 3). *Understanding the psychology and benefits of music therapy*. Save the Music Foundation. https://www.savethemusic.org/blog/music-therapy-and-mental-health/

Powers, W. (2003, October). *The piano: The pianofortes of Bartolomeo Cristofori (1655–1731).* The MET. https://www.metmuseum.org/toah/hd/cris/hd_cris.htm

Righetto, A. (2023, October 14). *Decoding the secrets of Mona Lisas mysterious smile: exploring theories, scientific analysis, and symbolism.* Mona Lisa's Daughter. https://www.monalisathedaughter.com/mona-lisa-painting/secrets/decoding-the-secrets-of-mona-lisas-mysterious-smile-exploring-theories-scientific-analysis-and-symbolism

Roald Amundsen: Polar explorer. (n.d.). Royal Museums Greenwich. https://www.rmg.co.uk/stories/topics/roald-amundsen-polar-explorer

Saiz-Jimenez, C. (2022, August 29). *Journey into darkness: Microbes living in caves and mines.* Frontiers for Young Minds. https://doi.org/10.3389/frym.2022.739199

Scalf, F. (n.d.). *The Rosetta Stone: Unlocking the ancient Egyptian language.* ARCE. https://arce.org/resource/rosetta-stone-unlocking-ancient-egyptian-language/

Solly, M. (2022, September 27). *Two hundred years ago, the Rosetta Stone unlocked the secrets of ancient Egypt.* Smithsonian Magazine.

https://www.smithsonianmag.com/history/rosetta-stone-hieroglyphs-champollion-decipherment-egypt-180980834/

Soto, C. (2022, August 29). *How has the Internet evolved?* Esferize.https://www.esferize.com/en/how-has-internet-evolved/

The story behind the Greenlandic tupilak. (2023, February 23). *Polar Quest.* https://www.polar-quest.com/blog/greenland/the-story-behind-the-greenlandic-tupilak

The strangest weather phenomena in the world. (2020, September 9). Love Exploring. https://www.loveexploring.com/gallerylist/99732/the-strangest-weather-phenomena-in-the-world

Tillman, N. T., & Harvey, A. (2024, March 5). *What are wormholes?* Space.com. https://www.space.com/20881-wormholes.html

Todd, P. (2023, October 19). *The lost city of Atlantis: Underwater wonders and myths.* Science Digest. https://sciencedigest.org/the-lost-city-of-atlantis-underwater-wonders-and-myths/

Tropical Cyclone Naming. (2019). World Meteorological Organization. https://community.wmo.int/en/tropical-cyclone-naming

Tsunami generation: Earthquakes. (n.d.). National Oceanic and Atmospheric Administration. https://www.noaa.gov/jetstream/tsunamis/tsunami-generation-earthquakes#:~:text=Earthquakes%20generally%20occur%20on%20three,from%20earthquakes%20on%20reverse%20faults.

The Tupilak: Mystical artifacts of Greenland's Inuit culture. (2024, August 24). *Dyane's Discoveries.*

https://daynesdiscoveries.com/2024/08/28/the-tupilak-mystical-artifacts-of-greenlands-inuit-culture/

What are dark matter and dark energy?. (2023, January 26). The Economist. https://www.economist.com/films/2023/01/26/what-are-dark-matter-and-dark-energy?utm_medium=cpc.adword.pd&utm_source=google&ppccampaignID=19495686130&ppcadID=&utm_campaign=a.22brand_pmax&utm_content=conversion.direct-response.anonymous&gad_source=1&gclid=Cj0KCQjwpvK4BhDUARIsADHt9sSyfjkVXykOtVkLAt0YNAbAj2a-H50lLcvYuFARuPqU7UCRRO-9MMsaAvr-EALw_wcB&gclsrc=aw.ds

What is an axolotl and why are they endangered?. (2024, April 2). Bluereef. https://www.bluereefaquarium.co.uk/portsmouth/blog/education/what-is-an-axolotl-and-why-are-they-endangered/

What is space weather and how does it affect the Earth? (n.d.). Center for Science Education. https://scied.ucar.edu/learning-zone/sun-space-weather/what-space-weather

Made in the USA
Las Vegas, NV
23 November 2024